VOLUME ONE HUNDRED AND SIX

ADVANCES IN GENETICS

Cosmic Genetic Evolution

ADVANCES IN GENETICS, VOLUME 106

Serial Editor

Dhavendra Kumar
William Harvey Research Institute
Bart's and The London School of Medicine & Dentistry
London, UK

VOLUME ONE HUNDRED AND SIX

ADVANCES IN GENETICS

Cosmic Genetic Evolution

Edited by

EDWARD J. STEELE
C.Y.O'Connor ERADE Village Foundation, Piara Waters, Perth, WA; Melville Analytics Pty Ltd, Melbourne, VIC, Australia Centre for Astrobiology, University of Ruhuna, Matara, Sri Lanka

N. CHANDRA WICKRAMASINGHE
University of Buckingham, Buckingham, United Kingdom; Centre for Astrobiology, University of Ruhuna, Matara; National Institute of Fundamental Studies, Kandy, Sri Lanka; Institute for the Study of Panspermia and Astrobiology, Gifu, Japan

ELSEVIER

ACADEMIC PRESS
An imprint of Elsevier

Academic Press is an imprint of Elsevier
50 Hampshire Street, 5th Floor, Cambridge, MA 02139, United States
525 B Street, Suite 1650, San Diego, CA 92101, United States
The Boulevard, Langford Lane, Kidlington, Oxford OX5 1GB, United Kingdom
125 London Wall, London, EC2Y 5AS, United Kingdom

First edition 2020

ISBN: 978-0-12-821518-0
ISSN: 0065-2660

For information on all Academic Press publications
visit our website at https://www.elsevier.com/books-and-journals

Publisher: Zoe Kruze
Editorial Project Manager: Hilal Johnson
Production Project Manager: Abdulla Sait
Cover Designer: Matthew Limbert

Typeset by SPi Global, India

Contents

Contributors

Dhammika Magana Arachchi
National Institute of Fundamental Studies, Kandy, Sri Lanka

Reginald M. Gorczynski
Department of Surgery & Immunology; University Toronto Health Network, Toronto General Hospital, University of Toronto, Toronto, ON, Canada

Brig Klyce
Astrobiology Research Trust, Memphis, TN, United States

Robyn A. Lindley
GMDx Group Ltd; Department of Clinical Pathology, Faculty of Medicine, Dentistry & Health Sciences, University of Melbourne, Melbourne, VIC, Australia

Yongsheng Liu
Henan Collaborative Innovation Center of Modern Biological Breeding, Henan Institute of Science and Technology, Xinxiang, China

Predrag Slijepcevic
Brunel University, London, United Kingdom

Willam A. Smith
Institute for the Study of Panspermia and Astroeconomics, Gifu, Japan

Edward J. Steele
Centre for Astrobiology, University of Ruhuna, Matara, Sri Lanka; C.Y.O'Connor ERADE Village Foundation, Piara Waters, Perth, WA; Melville Analytics Pty Ltd, Melbourne, VIC, Australia

Robert Temple
The History of Chinese Science and Culture Foundation, Conway Hall, London, United Kingdom

Gensuke Tokoro
Institute for the Study of Panspermia and Astroeconomics, Gifu, Japan

Dayal T. Wickramasinghe
College of Physical and Mathematical Sciences, Australian National University, Canberra, ACT, Australia

N. Chandra Wickramasinghe
University of Buckingham, Buckingham, United Kingdom; Centre for Astrobiology, University of Ruhuna, Matara; General Sir John Kotelawala Defence University, Ratmalana; National Institute of Fundamental Studies, Kandy, Sri Lanka; Institute for the Study of Panspermia and Astrobiology; Institute for the Study of Panspermia and Astroeconomics, Gifu, Japan

Foreword: Some things are simply given

Brig Klyce
Astrobiology Research Trust, Memphis, TN, United States

> *Major discoveries... are not like the discovery of America, where the general nature of the discovered object is already known. Rather, they are like recognizing that one has been dreaming*
>
> ***Paul Feyerabend (1995).***

Some things are simply given. Consider the planets, stars and galaxies, and the physical properties that make them possible. Whether our universe began in the finite past, or always existed, either way, it is unexplained by any prior natural cause. The existence of the physical world is simply given.

Now consider life. As we know, life comes from life. But we believe this series must have a beginning. Formerly we held that belief because there was no other way for life to arise on Earth. (Now we know there is another way, panspermia.) Still, we believe life must have originated at some time, because, by consensus, there was a big bang. If the whole universe began in a single, immeasurably hot explosion a finite time ago, logically, a subsequent origin of life from nonliving matter, somewhere in the universe, is required.

Interestingly, the big bang creation-story for the universe enjoys support from two opposing groups: advocates of the theory of evolution, and advocates of intelligent design (ID). By this powerful coalition, comprising at least 90% of interested, informed adults, the big bang theory is vigorously defended. We understand why IDers with religious leanings endorse this apparent miracle. But the reasoning from strict science eludes us. "Nothing comes from nothing."

Of course, the two groups diverge completely on the means for life's origin. IDers rely on unspecified supernatural agency (which also eludes us). Evolutionists rely on chemical reactions which cannot be demonstrated, nor even described in plausible detail. As our understanding of molecular biology and genetics improves, the apparent difficulties for a natural origin of life only grow. This conundrum has finally permitted panspermia to become respectable, because if life may have originated anywhere in the perceivable universe, the odds improve by twenty or more orders of magnitude.

But even that improvement does not make the math work out. Is there another approach to the problem? Stepping back, we notice that the origin of life is an assumption, not an observed phenomenon. Given the mounting difficulties, the assumption is increasingly burdensome. Only the consensus big bang requires life to originate. Could the big bang theory be flawed?

Of course it could. The theory is too new and to fluid to claim absolute certainty. And major surprises are not rare. For example, until 1998, by all expectations, the expansion of the universe was decelerating, slowed by gravity. Then we learned the opposite: the expansion is accelerating. In general, the cosmological data are loose, and are consistent with other scenarios that do not begin all at once from nothing. An informed minority of cosmologists actually do doubt the consensus big bang theory. The most highly honored living cosmologist, 2019 Nobel laureate James Peebles, is among them (Couronne & Ahmed, 2019).

If the origin of life is not logically required, perhaps life has always existed. In other words, maybe life, also, is simply given. Once that door is opened, it is reasonable to ask afresh: What is the tangible, observable evidence for the origin of life? We see very little. It reminds us of alchemy before the adoption of the periodic table of the elements.

If life has always existed, now evolution invites new scrutiny. How does it work? The easy answer is to say that once we have prokaryotic life, neo-Darwinism adequately explains everything that follows. But this ignores the actual evidence. The evidence supports neo-Darwinism only for optimizing existing programs within narrow ranges by point mutations—microevolution. Macroevolution, by contrast, requires whole new programs or subroutines. These have not been shown to arise by neo-Darwinian trial and error, in spite of hopeful, worldwide research. Instead, for macroevolution, exons, whole genes, and suites of genes must be properly acquired and activated. The final activation may require only a small mutation, but the original composing of these longer sequences remains unobserved and unexplained. Instead, new genetic programs seem to be *acquired*.

For just one example, In 2018, a team at Heinrich Heine University, Düsseldorf, wrote, *...we identify 3323 phenotypic innovations in the history of the E. coli clade that arose through changes in accessory genome content. ...Strikingly, every observed innovation arose through the horizontal acquisition of a single DNA segment less than 30 kb long. ...we found no evidence for the contribution of selectively neutral processes....* (Pang & Lercher, 2018). Horizontal gene transfer (HGT) is required.

If this is correct, we would expect to see evidence that genes exist before they are expressed. Indeed in 2003, Harvard's Andrew Knoll remarked that the genes underlying animal development always look older, by "molecular clock" dating, than the fossil record says they should be (Knoll, 2003). The genes appear to come first.

If genes come first, and are acquired by HGT, another assumption begins to arouse notice: that genetic programs ever *originated.* The evidence indicates only that they are *acquired.* Now we can wonder, as before: Perhaps they always existed. So when we suggest that life may have always existed, we mean all of it. If this possibility is accepted for consideration, we think the facts fit together much better. We understand that this view would represent a radical paradigm shift. But we think it may be time.

In fact, the suggestion is not new. A century ago, before the big bang was imagined, the President of the British Association for the Advancement of Science endorsed it. ...*We must begin seriously to consider whether the course of Evolution can at all reasonably be represented as an unpacking of an original complex which contained within itself the whole range of diversity which living things present.* (Bateson, 1914).

Under this philosophy, science would be free to undertake to explain everything that happens in the physical world, as deeply into the past as the evidence reaches. Nothing miraculous, or astronomically unlikely, needs to be invoked. Religious thinkers, on the other hand, are free to hold that things simply given are given by God. No hard evidence could prove otherwise. Consequently, grounds for conflict between science and religion would vanish. We think this is a happy outcome.

References

Bateson, W. (1914). *Presidential address at the Australian meeting of the British Association for the advancement of science, 1914.* http://www.archive.org/stream/reportofbritisha15adva/reportofbritisha15adva_djvu.txt.

Couronne, I., & Ahmed, I. (2019). *Top cosmologist's lonely battle against 'Big Bang' theory.* Agence France-Presse. https://phys.org/news/2019-11-cosmologist-lonely-big-theory.html.

Feyerabend, P. (1995). *Killing time: The autobiography of Paul Feyerabend.* Chicago University Press, p 92.

Knoll, A. (2003). *Life on a young planet: The first three billion years of evolution on earth.* ISBN: 0-691-12029-3, Princeton University Press, 202.

Pang, T. Y., & Lercher, M. J. (2018). Each of 3,323 metabolic innovations in the evolution of E. coli arose through the horizontal transfer of a single DNA segment. *Proceedings of the National Academy of Sciences of the United States of America, 116,* 187–192.

Preface: The origin of life and astrobiology

Charles Darwin is historically renowned for his theory of evolution as the basis of variation of biological life that exists around us. Similar views were also echoed by the contemporary biologist, Alfred Russel Wallace, who is credited with independently conceiving the theory of evolution through natural selection. His paper on the subject was jointly published with some of Charles Darwin's writings in 1858. Despite strong opposition of the contemporary Christian Church of the time, gradually the natural selection theory for the biological evolution was accepted and remains pivotal to explain both inter- and intraspecies variation. However, a small section of theologists, philosophers and biologists continue to hold different views for the origin of life and its developments into numerous species and variation. Both schools, however, believe and acknowledge the genetic basis of life within the realms of nucleic acids, the ribose nucleic acid (RNA) and the deoxyribose nucleic acid (DNA). The fundamental questions remain—what must have been the earliest form of life on Earth? What is the relationship to life beyond Earth, if it exists at all?

Molecular biologists agree that the earliest form of life must have been based on the RNA molecule, with DNA evolving later following successive chemical processes to facilitate the RNA functions. It is argued that the earliest life would have been heterotrophic, arising from and metabolically processing prebiotic organics as proposed in the Oparin-Haldane theory. In this context, it is important to consider the abiogenesis theory that life on Earth arose from nonlife more than 4 billion years ago. Abiogenesis proposes that the original life forms were very simple and gradually became increasingly complex. It probably preceded biogenesis, in which life is derived from the procreation of other life. Abiogenesis became impossible once the present composition of the Earth was achieved. The abiogenesis theory is neither proved nor disproved. The cosmic origins of life discussed in the many articles in this volume remains at the forefront of modern discussions.

In the 1920s, the famous British biometric scientist John Haldane and Aleksandr Oparin, the Russian biochemist independently set forth similar ideas for the origin of life on Earth. Both believed that organic molecules could be formed from abiogenic materials in the presence of an external energy source, for example the ultraviolet radiation, in combination with very low atmospheric oxygen, ammonia, water vapor, and probably other gases. Both agreed that the earliest form of life probably first appeared in the warm oceans of Earth and were heterotrophic, nurtured by preformed nutrients from chemicals in existence on early Earth, or brought in by comets compared to being autotrophic, generating food and nutrients from complex interactions of the sunlight, including cosmic radiations with inorganic materials. Oparin believed that the life developed from microscopic spontaneously formed spherical lipid molecules held together by electrostatic forces, probably the earliest form of cells. These molecules most likely functioned as enzymes, essential for the biochemical metabolic reactions necessary for life's evolution. Haldane, unfamiliar with Oparin's theory, believed that simple organic molecules formed first and in the presence of ultraviolet light became increasingly complex, ultimately forming cells. Haldane and Oparin's ideas formed the foundation for much of the research on the origin of life, specifically formation of cells.

The Haldane-Oparin theory on the origin of life was tested to a limited extent by two American chemists, Harold Urey and Stanley Miller. They successfully produced organic molecules from some of the inorganic components thought to have been necessary for the appearance of life, the prebiotic phase. The Miller-Urey experiment included combination of warm water with a mixture of water vapor, methane, ammonia, and molecular hydrogen exposed to the atmospheric electrical discharges in the form of lightening. Miller and Urey found that simple organic molecules, including amino acids (the building blocks of peptides), had formed under the simulated conditions of early Earth. The Miller-Urey experiment demonstrated that organic molecules could form from abiogenic materials under the constraints of Earth's prebiotic atmosphere. Later research showed that amino acids can spontaneously form small protein molecules (peptides). It is also shown that the RNA molecule can be artificially synthesized from nucleotides (nitrogen containing compounds or bases) linked to sugar and phosphate groups. In fact, in addition to carrying and translating genetic information, RNA acts like a catalyst, a molecule that increases the rate of a reaction without itself being consumed. It is, thus, logical to accept that multiple forms of RNA existed in the prebiotic phase of abiogenesis that led to the formation of life on Earth or in any wider context.

It has been suggested that peptide nucleic acid (PNA) might have been the first genetic material that preceded the RNA and DNA. PNA forms very stable double helical structures and even stable triple helices. However, the chemical nature of PNA does not allow it be a sustainable genetic material necessary for replication, transfer of genetic information or its reorganization. Activated PNA monomers tend to cyclize easily and thus formation of oligomers is very difficult under prebiotic conditions. Further, and PNA hydrolyses rather rapidly and thus restricts the chances of it ever being accumulated in sufficient quantity, for instance deep in the primitive oceans.

The concept of sunlight (ultraviolet radiation) and energy sources from other planets in the cosmos (cosmic radiations) influencing the origin of life on Earth, or indeed anywhere in the cosmos, falls within the field of Astrobiology. The main question remains on the likelihood of the existence of life beyond the confines of Earth or perhaps in the wider cosmos. Astrobiology deals with the processes that may lead to the origin and evolution of life on Earth. Scientists continue to focus first on sites for understanding the processes of chemical (prebiotic) evolution (Titan, asteroids, and comets) and second on targets for searching for life beyond Earth (Mars, Venus, Europa, Enceladus, and others). Mars is one of the most popular targets due to the fact that early Mars and early Earth had a similar history. The quest for life on Mars is one of the most important missions of the modern astrobiology. Several ethicists and philosophers have strong reservations in relation to ethical, social and legal concerns that fall within the new discipline of *astrobioethics*. Perhaps the main stumbling block of research in life on Mars is the anticipation of a close similarity to life on Earth.

Most astrobiologists are guided by the following broad requirements and principles on the origin of life:

(1) The elements of life, carbon, hydrogen, oxygen, and nitrogen (CHON).
(2) CHON + energy (e.g., UV from stars, atmospheric lightning discharge) = monomers (Miller-Urey micromolecules).
(3) Monomers + aqueous environment (to protect prebiotic organics from destruction by UV) + concentration/assembly/ordering = polymers.
(4) Some such polymers presumably incorporated both gene-like information and enzyme-like catalytic activity (ribozymes). The information content in the enzymes being of a super-astronomical magnitude opens the way to the concept of a cosmological origin of life and the ideas of panspermia discussed by Sir Fred Hoyle and Prof. Chandra Wickramasinghe.
(5) Earliest life, wherever it started, would thus have been heterotrophic and thermophilic.

The concept of cosmic genetic evolution emerged from discussions with the guest editors of this thematic volume. It is an extension of astrobiology to explore the extent of ultraviolet radiation (sunlight) and range of complex cosmic radiation systems implicated in genetic or genomic variation and function. The cosmic radiations (cosmic rays) include high-energy protons and atomic particles (nuclei) which move through space at nearly the speed of light. They originate from the sun, from outside of the solar system, supernova explosions of star, and from distant galaxies. These pass through the ozone layer and upon impact with the Earth's atmosphere, cosmic rays can produce cascade of secondary particles capable of striking the Earth's surface.

The fundamental questions remain:

- Whether cosmic rays are capable of inducing spontaneous benign or deleterious genetic changes (mutations and variants)?
- How far the cosmic ray phenomenon along with the introduction of new cosmic viruses could account for successive evolution of life on Earth?
- Could cosmic radiation be an important factor in the emergence of different forms of biological life and variations?
- How far does the impact of cosmic rays on the Earth might account for the creation of new viral variants capable of causing human disease, for example the Corona/Covid-19/SARS-CoV-2 virus?

The guest editors and other contributing authors have produced this impressive volume on a very unusual subject full of past and present hypotheses. Information and supporting arguments are most relevant and will form the basis for further discussion and design of new experiments and investigations. They all deserve to be congratulated on this exciting achievement. Now authors, editors and publishers shall wait for acceptance and appreciation of this unique contribution to the genetics literature.

Disclaimer and acknowledgements

This article is based on author's own perceptions, understanding and amateur knowledge of cosmology and astrobiology. Some of the ideas and concepts are derived from author's informal discussions with Professor Chandra Wickramasinghe, the Co-Editor of this thematic volume. His thoughtful insights and guidance are greatly appreciated. The author prefers to avoid add literature references as the subject is widely discussed in diverse digital open access public domains.

The author would be delighted to deal with any questions or observations related to this article: d.kumar@qmul.ac.uk; genomicmedicineuk@gmail.com.

Dhavendra Kumar
Edward J. Steele
N. Chandra Wickramasinghe

CHAPTER ONE

Introduction—Panspermia, 2020

Edward J. Steele*
C.Y.O'Connor ERADE Village Foundation, Piara Waters, Perth, WA, Australia
Centre for Astrobiology, University of Ruhuna, Matara, Sri Lanka
Melville Analytics Pty Ltd, Melbourne, VIC, Australia
*Corresponding author: e-mail address: e.j.steele@bigpond.com

Abstract

This current volume is, in many ways, a 2020 update to the important 1999–2000 compendium by Sir Fred Hoyle and Professor N. Chandra Wickramasinghe's "Astronomical Origins of life: Steps towards Panspermia." The emerging new paradigm of biology that connects life on Earth with the wider cosmos is covered in considerable depth showing that terrestrial biological evolution is best understood as a cosmically derived habitat and an interconnected genetic system. The various chapters here discuss all aspects of this interconnectedness, particularly relevant now in this time of the coronavirus pandemic (COVID-19) as the human race reacts to the many microbes and viral pathogens that arrive regularly from space.

It is now 50 years since the modern revival of scientific investigations into astrobiology began with the work of the late Sir Fred Hoyle and Professor N. Chandra Wickramasinghe on defining astrobiology as a universe-wide phenomenon (Panspermia). The work of Hoyle-Wickramasinghe and their numerous colleagues marshaled the first key evidences and supplied the critical scientific analyses and arguments to allow the interpretation of the emerging new data. These new data relate directly and indirectly (via "biosignatures" and the like) to extra-terrestrial biological seedings events which then influenced the ongoing terrestrial evolution of life on Earth (Steele et al., 2018). Indeed the current volume can be considered an update to their important compendium "Astronomical Origins of life: Steps towards Panspermia" (Hoyle & Wickramasinghe, 2000). In that compendium the Preface was entitled "Panspermia 2000" and we use this same updated title to flag the appropriate contents of the present book.

This volume therefore brings together various aspects of an emerging new paradigm of biology that connects life on Earth with the wider cosmos. The available facts of terrestrial biological evolution we believe can

Advances in Genetics, Volume 106
ISSN 0065-2660
https://doi.org/10.1016/bs.adgen.2020.04.001

only be understood properly on the basis of a cosmically derived and interconnected genetic system. Indeed as various chapters here discuss our very well being appears overtly dependent on how human beings react biologically to the numerous microorganisms that arrive regularly from space (Steele et al., "Origin of new emergent Coronavirus and Candida fungal diseases—Terrestrial or cosmic?," this volume).

The picture that now emerges is that life on Earth was been seeded from space and began to flourish here at, or shortly after, the *Hadean* epoch about 4.1 billion years ago (Wickramasinghe et al., "Cometary panspermia and origin of life?," this volume). The ongoing and varied tempo and type of these continuing "in-falls" of cosmic genetic material contributed to the emergence and continued evolution of diverse living systems here on Earth. Without such a connection with the wider cosmos we argue that no significant evolution of life would have been possible. It is thus now apparent that the traditional slow, blind and random evolutionary mechanisms, assumed by the traditional neo-Darwinian theory of evolution, are inadequate to explain a wide range of new astrophysical and biological data. Neo-Darwinism as the *dominant* evolutionary process is now contradicted on multiple fronts by the sheer weight and variety of the emerging new evidence. A switch to a Lamarckian-Panspermic view of life, is, in our view, long overdue. Some of these rapid genetic adaptation mechanisms are discussed in various Chapters (Steele et al., "The efficient Lamarckian spread of life in the cosmos," this volume; Lindley, "The mutagenic source and power of our own evolution," this volume; Steele et al., "Origin of new emergent Coronavirus and Candida fungal diseases—Terrestrial or cosmic?," this volume).

Since the 1970s a range of cross-disciplinary evidence has emerged for Lamarckian genetic processes of acquired inheritance—rapid, fast and directional changes that are environmentally responsive. That is, the genetic changes we attribute to evolution of organisms are sensitive to *genetic feedback mechanisms* that guide their future evolutionary development. In lower organisms this is manifest as 'adaptive evolution' where single cells (prokaryote, archaea, more complex eukaryotes) employ rapid SOS or hypermutation and rapid selection proliferation strategies—to ensure survival in new environments. In higher multicellular animals adaptive somatic selection strategies involving acquisition of new genes both by horizontal gene transfers HGT), coupled with somatic hypermutation, reverse transcription and rapid somatic selection allow "adapted" progeny somatic cells and new vertically transmitted germline genes to emerge (viz. the penetration of the Soma-Germline *Weismann Barrier*). Plants do not possess a *Weismann Barrier* but use a similar set of Lamarckian strategies to pass on their acquired adaptive somatic genetic solutions to progeny organisms, in a process aptly identified as a Pangenesis process, exactly in principle as envisaged by Charles Darwin himself 150 years ago (Liu, 2018).

So these diverse new emerging observations and data on "acquired inheritance" across the kingdoms of life now make rational sense in terms of Lamarckian modes of inheritance (Steele et al., "The efficient Lamarckian spread of life in the cosmos," this volume). We thus draw out the scientific links between rapid Lamarckian inheritance and the efficient Panspermic spread of pre-existing living systems throughout the Cosmos (Steele et al., 2019). This can be summed up as follows: A population of living cellular systems—spores, seeds, even cryopreserved fertilized eggs encased within protective matrices (e.g. interiors of wandering comets, small moons, planetoids, or fragments thereof)—may be dispersed throughout the Cosmos via long space-time journeys (hundreds of millions to billions of years). On landing in a new Cosmic niche a portion or subset of the cells surviving impact adapt rapidly to their new environment primarily via Lamarckian rapid adaptation genetic mechanisms (which may often cause overt disease, Hoyle & Wickramasinghe, 1979, Steele et al., "Origin of new emergent Coronavirus and Candida fungal diseases—Terrestrial or cosmic?," this volume). Thus a Panspermic universe provides the scientific rationale for the actual widespread existence of Lamarckian modes of genetic evolution both on Earth and throughout the Cosmos.

What then for the actual origin of life itself? Our pragmatic position is summed up recently (Steele et al., 2019) but done succinctly here in the Foreword by Brig Klyce who has spent more than 25 years curating the relevant information and the key discoveries at the Cosmic Ancestry website (www.panspermia.org).

References

Hoyle, F., & Wickramasinghe, N. C. (1979). *Diseases from space*. London: J.M. Dent Ltd.

Hoyle, F., & Wickramasinghe, N. C. (2000). Astronomical origins of life: Steps towards Panspermia. Panspermia 2000. Preface to Reprints from Astrophys. Space Sci, In (Vol. 268). Dordrecht, The Netherlands: Kluwer Academic Publishers, pp. 1–3, 1999.

Liu, Y. (2018). Darwin's pangenesis and graft hybridization. *Advances in Genetics*, *102*, 27–66.
Steele, E. J., Al-Mufti, S., Augustyn, K. A., Chandrajith, R., Coghlan, J. P., Coulson, S. G., et al. (2018). Cause of Cambrian explosion—Terrestrial or cosmic? *Progress in Biophysics and Molecular Biology*, *136*, 3–23.
Steele, E. J., Gorczynski, R. M., Lindley, R. A., Liu, Y., Temple, R., Tokoro, G., Wickramsinghe, D. T., & Wickramasinghe, N. C. (2019). Lamarck and Panspermia—On the efficient spread of living systems throughout the cosmos. *Progress in Biophysics and Molecular Biology*, *149*, 10–32.

CHAPTER TWO

Cometary panspermia and origin of life?

N. Chandra Wickramasinghe[a,b,c,d,*], Dayal T. Wickramasinghe[e], and Edward J. Steele[b,f,g,*]
[a]University of Buckingham, Buckingham, United Kingdom
[b]Centre for Astrobiology, University of Ruhuna, Matara, Sri Lanka
[c]Institute for the Study of Panspermia and Astrobiology, Gifu, Japan
[d]National Institute of Fundamental Studies, Kandy, Sri Lanka
[e]College of Physical and Mathematical Sciences, Australian National University, Canberra, ACT, Australia
[f]C.Y.O'Connor ERADE Village Foundation, Piara Waters, Perth, WA, Australia
[g]Melville Analytics Pty Ltd, Melbourne, VIC, Australia
[*]Corresponding authors: e-mail address: ncwick@gmail.com; e.j.steele@bigpond.com

Contents

Abstract

A range of astronomical observations are shown to be in accord with the theory of cometary panspermia. This theory posits that comets harbor a viable biological component in the form of bacteria and viruses that led to origin and evolution of life on Earth. The data includes (1) infrared, visual and ultraviolet spectra of interstellar dust, (2) infrared spectra of the dust released from comet Halley in 1986, (3) infrared spectra of comet Hale-Bopp in 1997, (4) near and mid-infrared spectra of comet Tempel I in 2005, (5) the discovery of an amino acid and degradation products attributable to biology in the material recovered from the Stardust Mission in 2009, (6) jets from comet Lovejoy showing both a sugar and Ethyl alcohol and finally, (7) a diverse set of data that has emerged from the Rosetta mission. The conjunction of all the available data points to cometary biology and interstellar panspermia as being inevitable.

Advances in Genetics, Volume 106
ISSN 0065-2660
https://doi.org/10.1016/bs.adgen.2020.03.003

1. Historical context

The pre-Socratic Greek philosopher Anaxagoras (510–428 BCE) was probably the first to introduce the idea of panspermia (pans = everywhere, spermata = seeds–seeds everywhere) into Western Philosophy. Despite the cogency of Anaxagoras' arguments for panspermia the rival theory of spontaneous generation that was introduced much later by Aristotle (384–322 BCE) has remained overwhelmingly the dominant point of view well into the 19th century. Spontaneous Generation asserts life emerges from non-living material whenever the right conditions prevail—with the famous metaphor of fireflies emerging from a mixture of warm earth and morning dew! Over the centuries many challenges to spontaneous generation have emerged, but none that gained any significant traction until the latter half of the 19th century.

Louis Pasteur's classic experiments in 1860 on the processes of fermentation—brewing of wine and souring of milk—demonstrated convincingly that microbial life appears always to be derived from pre-existing microbial life. On the basis of such experiments he famously declared that "spontaneous generation was finally dead" and also most memorably enunciated the dictum "*omne vivum e vivo*"—all life from life. But of course life is never as easy as that!

Pasteur's classic experiments led to an immediate revival of interest in panspermia as a possible mode of origin for terrestrial life. In 1861, Lord Kelvin in his presidential address to the British Association stated thus:

> *"Hence, and because we all confidently believe that there are at present, and have been from time immemorial, many worlds of life besides our own, we must regard it as probable in the highest degree that there are countless seed-bearing meteoritic stones moving about through space. If at the present instant, no life existed upon the Earth, one such stone falling upon it might, by what we blindly call natural causes lead to its being covered with vegetation."*

Two decades later, the concept of panspermia was further elaborated upon by Nobel Laureate Svante Arrhenius, first in a short paper in 1903, and later in his 1908 book "Worlds in the Making." Many sceptics came forward in the early decades of the 20th century challenging Arrhenius with the assertion that bacterial life cannot survive the harsh conditions prevailing in space.

However, many decades before the discovery of extremophiles and of the radiation resistance of bacteria, Svante Arrhenius had already inferred that such survival properties must exist from experiments he himself had conducted by taking seeds down to near 0 K and finding them to survive. Arrhenius also calculated correctly the effect of radiation pressure of starlight on spores and argued that such living particles (bacteria and spores) can be propelled quite naturally from one star system to another by this means. Thus the concept of panspermia was put into firm scientific setting.

Pasteur's experiments and his dictum—*Omne vivum e vivo*—remains the most powerful *raison d' ˆetre* for panspermia. If life is always derived from life in a causal chain, as is clearly manifest in present day life and also throughout the fossil record, the question naturally arises as to when and where this connection ceased. The continuation of the life-from-life chain to a time before the first life appears on our planet and before the Earth itself formed can be interpreted as implying that life is in essence as old as the Universe itself.

2. Earliest life on Earth

Three decades ago the earliest evidence for microbial life on the Earth was thought to be in the form of cyanobacteria-like fossils dating back to 3.5 Ga ago. From the time of formation of a stable crust on the Earth 4.3 Ga ago following an episode of violent impacts with comets (the Hadean Epoch) there then seemed to be available a comfortable 800 million years timespan during which the canonical Haldane-Oparin primordial soup may have conceivably developed. Very recent discoveries, however, have shown that this time interval has been effectively closed. Detrital zircons older than 4.1 Gy, discovered in rocks belonging to a geological outcrop in the Jack Hills region of Western Australia, have been found to contain micron-sized graphite spheres with an isotopic signature of biogenic carbon (Bell, Boehnke, Harrison, & Mao, 2015). The ^{12}C-enrichment found within these inclusions may thus be taken as unequivocal evidence for the existence of microbial life on Earth before 4.2 Billion years ago, during the epoch of comet and asteroid impacts. The requirement now, on the basis of orthodox thinking, is that an essentially instantaneous transformation of non-living organic matter to bacterial life took place, a fact that strains credibility of a possible Earth-bound abiogenesis beyond the limit. A far more plausible possibility is that fully-developed microorganisms arrived at the Earth via impacting comets, and these then became carbonized and trapped within condensing mineral grain conglomerates.

3. Improbability of the origin of life

With a lingering scientific obsession with the idea of Earth-bound spontaneous generation, the success of the famous Miller-Urey experiment of 1958/59 showing how organic molecules could be synthesized from mixtures of inorganic gases under laboratory conditions led to the conviction that it was only a matter of time before the next steps from biochemical monomers to life itself could be demonstrated in the laboratory (Miller & Urey, 1959). Despite over half a century of sustained effort this goal has proved stubbornly elusive (Deamer, 2011).

Starting from an initial small bacterium (typified for example by *Mycoplasma genitalium*) with ~500 genes, life apparently evolves over a 4 billion year timescale to produce the entire spectrum of terrestrial life including mammals and humans with genomes consisting of some 25,000 genes. Hoyle and one of the present authors have dwelt at length on the improbability of obtaining even a minimal gene set needed for the emergence of a bacterial genome from random processes in any restricted terrestrial-type setting. For a set of 500 genes, assuming that about 10 sites per gene need to be correctly filled with one of a set of 20 amino acids, the probability turns out to be $\sim 10^{-6500}$—a hurdle of truly cosmic proportions and one that is decisively insurmountable (and reviewed again for a minimal cell with 256 protein-coding genes in Steele et al., 2019).

If one accepts this calculation showing a grotesquely small a priori probability for the transition from non-life to life it would appear that only two options remain open. The origin of life on Earth was an extremely improbable event that somehow occurred (because we are here!) but will effectively never be reproduced again either here or elsewhere. Alternatively, an essentially infinite system of truly cosmological proportions as well as an immeasurably longer timescale was involved in an initial origination event. How big or how old that system needs to be is still a matter for debate. Arguments by Abel and Trevors (2006) and Abel (2009) suggest that within the framework of Big-Bang type cosmologies naturalistic protogene formation still faces almost insuperable difficulties. However, by whatever process life has emerged, this event of origination must be reckoned as unique, and the subsequent spread of life throughout the universe ensured by the processes of "panspermia."

4. Interstellar dust, polycyclic aromatic hydrocarbons (PAH) and biological molecules

Identifying the composition of dust in interstellar clouds led to a major incentive for embarking on the theory of cosmic life (Wickramasinghe, 1967). The interstellar dust absorbs and scatters starlight causing extinction or dimming of the light from stars, re-emitting the absorbed radiation in the infrared. Extinction of visual light shows up in the form of gigantic dark clouds and striations against the background of stars in the Milky Way. The total mass of dust amounts to a staggering few percent of the mass of all the stars.

An important clue relating to interstellar dust composition followed quite early from precise quantitative studies of the extinction (or dimming) of starlight. In the 1960s, from a combination of new astronomical observations and theoretical modeling it transpired that the reigning paradigm of the 1960s that the dust was largely comprised of water-ice was deeply flawed. New infrared observations showed the omnipresence of absorptions due to CH, OH, C–O–C linkages consistent with organic polymers (Hoyle, Wickramasinghe, & Al-Mufti, 1982; Wickramasinghe, 1974). Today the best overall agreement over the entire profile of interstellar extinction from the infrared to the ultraviolet demands in our view a mixture of desiccated bacteria, nanobacteria, including biologically derived aromatic molecules (Hoyle & Wickramasinghe, 1991, 2000).

A crucial turning point in the attempts to identify the precise nature of cosmic dust and compare it with biology came in the early 1980s with the infrared observations that had been requested by Hoyle and Wickramasinghe (1981). The infrared spectrum of a source known as GC-IRS7 (near the center of the galaxy) over the 2–4 μm waveband showed an absorption profile that matched biology to an amazing degree of precision. These data, following age-old standard procedures in Astronomy, were predicted by Hoyle and Wickramasinghe *in advance* of the interstellar dust observations by D.T. Wickramasinghe and Allen, as shown in Fig. 1. This figure shows the normalized IR extinction (absorption) flux for two independent data sets—astronomical observations and the IR absorption spectrum in the wavelength range 2.9–4.0 (μm) for desiccated (freeze dried) *E. coli* bacterial cells (solid line). This is an intricate

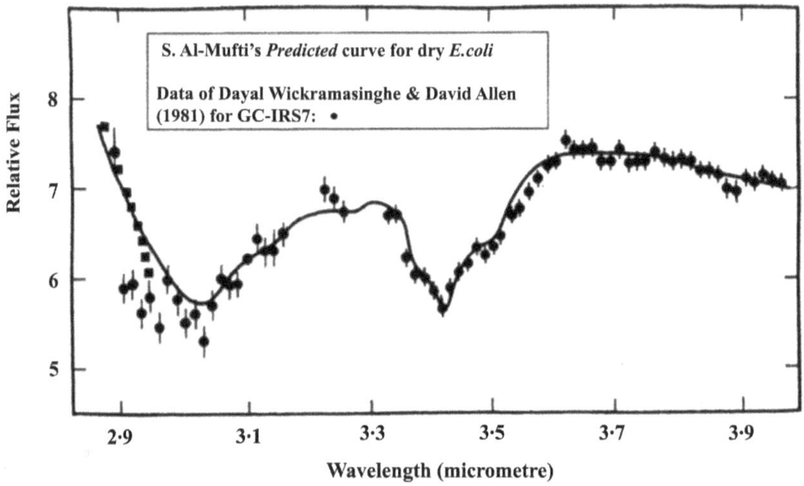

Fig. 1 Comparison of the infrared flux (arbitrary units) from the astronomical source GC-IRS7 near the galactic center, with the curve predicted for freeze dried *E. coli* cells (Allen & Wickramasinghe, 1981). This figure has also been reproduced in recent open access articles (Steele et al., 2018, 2019).

and complex IR absorption spectrum of living, albeit dried and dormant, living cells. The observational data points were secured at each wavelength indicated for IR electromagnetic radiation emitted from a source located some 23,000 light years away near the center of the Milky Way. As this IR light traverses through clouds of the dust grains it is absorbed in a similar fashion to the IR absorption by dry *E. coli* cells in the laboratory.

The data shown in Fig. 2, was confirmed independently by the team of Okuda et al. (1990). The Pearson correlation of this paired comparison data gives r as 0.9324 for $N=77$ pairs. For Okuda et al. (1990) the r value is 0.9275 for $N=35$ pairs. The P values for both are $<10^{-9}$. That is, we would expect to see such an exact spectral match by chance alone in more than 1 billion similar trials (DT Wickramasinghe, G Briggs, NC Wickramasinghe, EJ Steele, Unpublished calculations).

The simplest conclusion to be drawn from this correspondence is that the dust grains in the interstellar medium have infrared absorption properties over the entire continuum from 2.8 to 4.0 μm that are identical to desiccated bacterial cells. After 1983 many other spectral features of interstellar dust over other wavebands have also found ready explanations in terms of dust particles of biological origin. We note in particular that similar IR absorption spectral matches of larger eukaryotic cells such as diatoms (algae) were

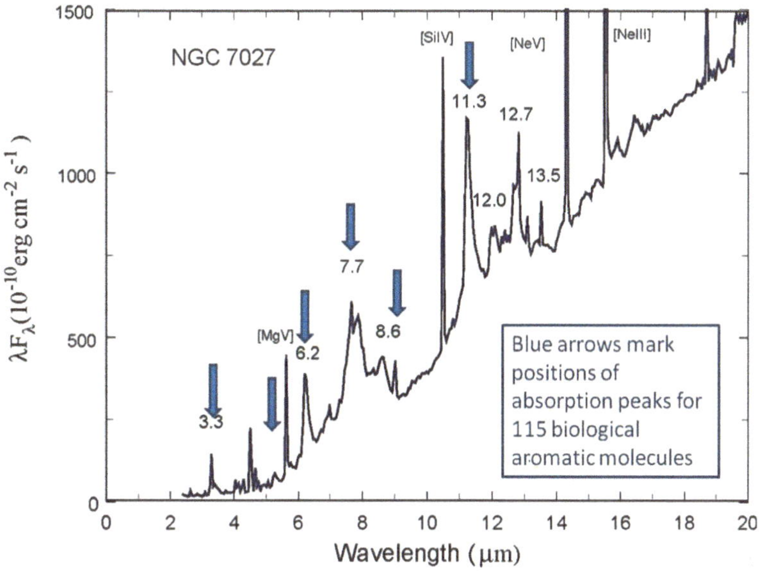

Fig. 2 IR spectrum of planetary nebula NCG7027 compared with main absorptions of 115 biological aromatic molecules (redrawn from data *in the public domain* obtained by the European Space Agency, Infrared Space Observatory).

obtained for the 8–13 um infrared wavelength range (e.g., Hoyle et al., 1982, Hoover, Hoyle, Wickramasinghe, Hoover, & Al-Mufti, 1986). To the best of our knowledge no artificial modeling of compound organic mixtures will produce such invariant and exact matches to astronomical data with any reasonable set of assumptions. It is difficult therefore to provide an interpretation for these data that avoids Panspermia.

There are also many other astronomical datasets that have been found to be consistent with panspermia. The distribution of the so-called unidentified infrared bands (UIB's) in stellar spectra between 3.3 and 22 μm is almost identical in their wavelengths in widely different astronomical sources, more or less irrespective of the ambient conditions (Rauf & Wickramasinghe, 2009). Data on UIB's for a large number of galactic and extragalactic sources have been obtained using the Spitzer Space Telescope and other orbiting telescopes. Fig. 2 shows the spectrum of the planetary nebula NGC 7027 with arrows pointing to the strong absorption bands in an aromatic mixture of chemicals derived from degraded biology.

While PAHs (Polyaromatic Hydrocarbons), often *presumed* without proof to form inorganically, are the favored model for the UIBs, no really satisfactory agreement with available astronomical data has thus far been

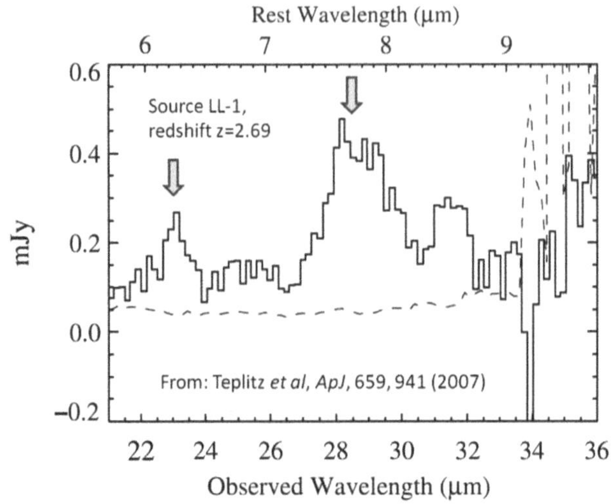

Fig. 3 Redshifted 6.2, 8.7, 11.3 μm biological bands in a source that emitted this radiation when the universe was 2 billion years old (Teplitz et al., 2007) (redrawn from published data).

shown possible for abiotic PAHs (Hoyle & Wickramsinghe, 1991; Rauf & Wickramasinghe, 2009).

Among the most distant galaxies displaying aromatic/biomolecular infrared signatures is a very distant high red-shift infrared luminous galaxy at redshift z = 2.69, the spectrum of which is shown in Fig. 3 (Teplitz et al., 2007). This galaxy emitted its light when the Universe was at the tender age of 2 billion years according to standard Big Bang cosmology.

Evidence of PAH-type biodegradation products also shows up in the form of a 2175A spectral signature in the ultraviolet in recent observations of the most distant galaxies up to redshifts of at least 2.45. We thus have clear evidence that biologically related materials were formed within 2.5 billion years of the presumed Big-Bang origin of the universe which is currently believed to be 13.8 billion years ago (Motta et al., 2002; Noterdaeme, Petitjean, Ledoux, & Srianand, 2009). The UV spectral feature for two intermediate redshift galaxies is shown in Fig. 4 compared with our calculated absorption profile for the biological aromatic ensemble that gives agreements with the IR and the mid UV absorptions in our galaxy (Hoyle & Wickramsinghe, 1991).

Detections of interstellar organic molecules of ever-increasing complexity including molecules relevant to biology have continued in the last two decades with the deployment of newer and better instruments and

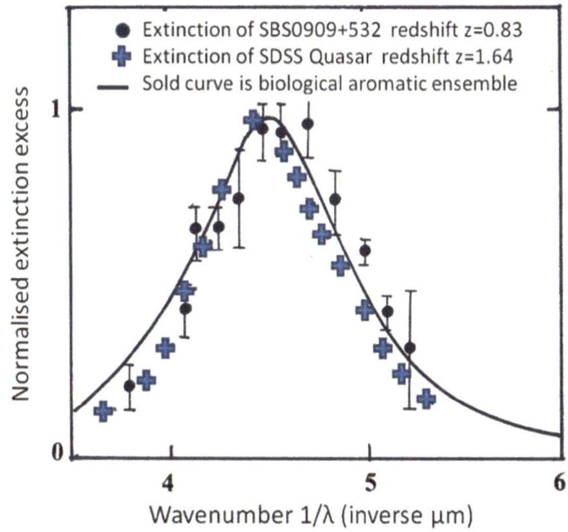

Fig. 4 The 2175A feature in two high redshift galaxies, compared with aromatic biomolecules (Motta et al., 2002; Noterdaeme et al., 2009) (redrawn from published data).

telescopes. Infrared, microwave and radio observations are used to detect the presence of such molecules and the current list of positive detections is likely to be constrained only by limitations of techniques.

5. Comets and biomaterial

The organic composition of comets and the concept of cometary panspermia was proposed and developed by Hoyle and Wickramasinghe (1981, 1985). The first observational evidence that had a direct bearing on the possible existence of cometary biology (rather than just organics) came with the 1986 perihelion passage of Comet Halley. The mid-infrared spectrum (Fig. 5) of the comet's dust coma following a major outburst on March 31, 1986 showed unambiguous evidence of aromatic-aliphatic linkages (C–H stretching modes) that were consistent with desiccated *E. coli* (Wickramasinghe & Allen, 1986, Wickramasinghe, Hoyle, Wickramasinghe, & Al-Mufti, 1986).

Since 1986 many comets have been studied both with ground based instruments as well as by techniques using spacecraft. All these studies have yielded generally similar results. Comets, in addition to water ice and silicate minerals have a large component of organic material that could most plausibly be explained in terms of biology and biological processing.

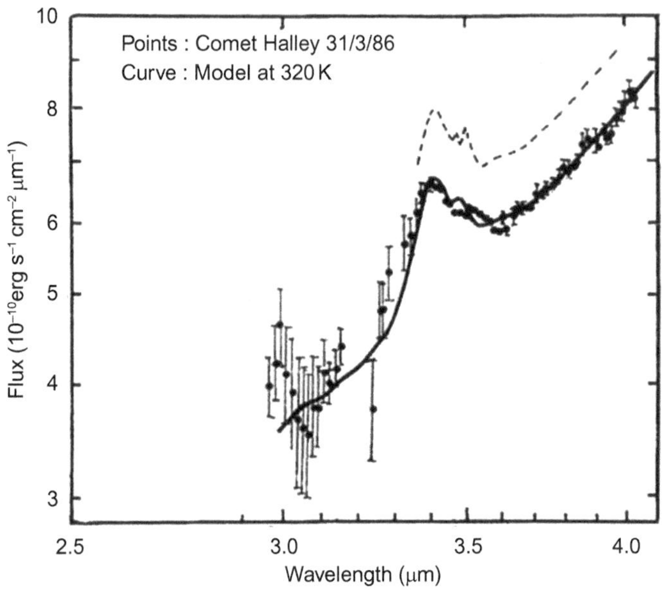

Fig. 5 Emission by dust coma of Comet Halley observed by Wickramasinghe and Allen (1986) (points) compared with normalized fluxes for desiccated *E. coli* at an emission temperature of 320 K. The solid curve is for unirradiated bacteria; the dashed curve is for X-ray irradiated bacteria (see Hoyle & Wickramasinghe, 2000, and reviewed in Steele et al., 2018).

Most recently the European Space Agency's Rosetta Mission to comet 67P/C-G has yielded the most detailed observations that satisfy consistency checks for biology and our ideas of cometary panspermia. Fig. 6 shows the close consistency between the surface properties of the comet and the spectrum of a desiccated bacterial sample.

Jets of water and organics were observed issuing from ruptures and vents in the frozen surface, again consistent with biological activity occurring within sub-surface liquid pools (Wallis & Wickramasinghe, 2015). The most recent report of O_2 along with evidence for the occurrence of water and organics provides further evidence of such ongoing biological activity (Bieler et al., 2015). Such a mixture of gases cannot be produced under thermodynamic conditions, because organics are readily destroyed in an oxidizing environment. The freezing of an initial mixture of compounds, including O_2, not in thermochemical equilibrium, has been proposed, but there is no evidence to support such a claim. On the other hand the oxygen/water/organic outflow from the comet can be explained on the basis of subsurface microbiology. Photosynthetic microorganisms operating at the low light levels near the surface at perihelion could produce O_2 along with organics.

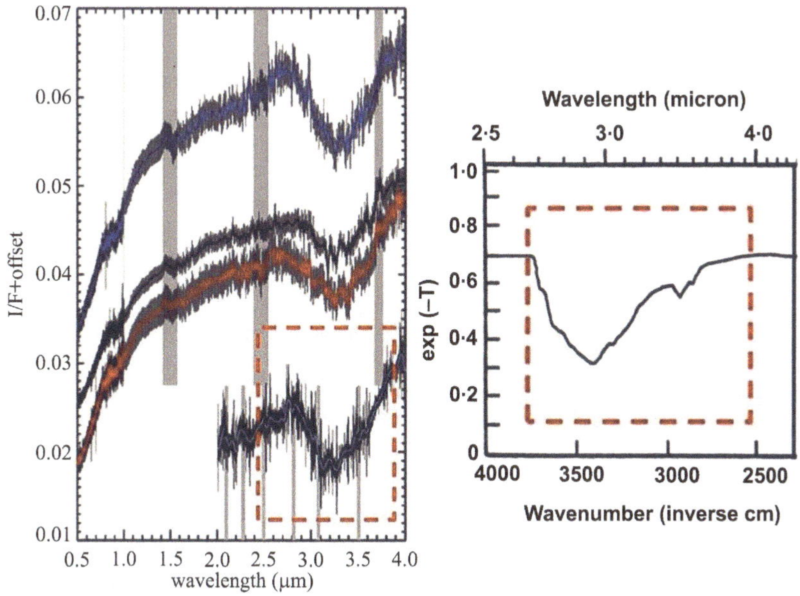

Fig. 6 The surface reflectivity spectra of comet 67P/C-G (left panel) compared with the transmittance curve measured for *E. coli* (right panel) (Capaccione et al., 2015, and reviewed again in Steele et al., 2019).

In our view outbursts of comet 67P/C-G could not result from the volatilization of the comet's surface, but from gas pressure built up in sub-surface (presumed liquid) domains. The contents of a pressurized 10^6–10^7 t lake flooding out with say 10% going as gases and entrained particles into the coma, could explain the outbursts of dust observed in comets (Wickramasinghe, Wainwright, Smith, et al., 2015). Although CO, CO_2 or CH_4 may well be constituents of the driver gas at the base of ruptured vents, most of this material would recondense on grains close to the surface while liquid water would form an ice crust due to sublimation cooling. The CO production (observed in comet Hale-Bopp) would most naturally be explained as a photodissociation product of volatile and fragile biochemicals that contain weakly bonded CO groups which re-evaporate from superheated smaller grains.

The recent report of O_2 along with evidence for the occurrence of water and organics in the same locations provides further evidence of such ongoing biological activity. Such a mixture of gases cannot be produced under thermodynamic conditions, because organics are readily destroyed in an oxidizing environment. The freezing of an initial mixture of compounds, including O_2, not in thermochemical equilibrium, has been proposed, but there is no evidence to support such a claim. On the other hand the oxygen/water/organic outflow from the comet can be readily explained on the basis of subsurface microbiology. Photosynthetic microorganisms operating at the low light levels near the surface at perihelion could produce O_2 along with organics. The combination of oxygen O_2, O_3 (oxidizing) with gases such as methane (reducing) is not permitted by thermodynamics so their coincidence in space may be regarded as *prima facie* evidence of biology.

Altwegg et al. (2016) have reported the discovery of the biological amino acid glycine and as well as an over-abundance of phosphorus in the coma of comet 67P/CG. Very high ratios of $P/C \sim 10^{-2}$ that have been observed are difficult to reconcile on the basis of volatilization of condensed inorganic material of solar composition, which have a $P/C \sim 10^{-3}$, particularly as we might expect *inorganic* phosphorus to be mostly fixed in refractory minerals. On the other hand, the P/C ratio of biomaterial is close to the value implied by the cometary data and would be explained if the material in the comet's coma started off as biomaterial—virions and bacteria (Fagerbakke, Heldal, & Norland, 1996). The idea that microbial life cannot exist in comets still dominates scientific orthodoxy but its rational basis is fast eroding (Wickramasinghe, 2011; Wickramasinghe & Tokoro, 2014). Adhering to a paradigm that has long past its sell-by date will be to the detriment of science.

Methyl chloride (CH_3Cl) and other haloalkanes are a significant source of atmospheric carbon via volatile haloalkanes in the terrestrial case and is mostly produced by oceanic biochemical processes via various macroalgal and other biomasses in the presence or absence of sunlight (Laturnus, 2001; Laturnus, Adaams, & Wiencke, 1998; Lobert, Keene, Logan, & Yevich, 1999). Further emission of CH_3Cl by wood rotting fungi contribute to some of the atmospheric complement of this molecule (Watling & Harper, 1998). The presence of significant amounts of this molecule and related haloalkanes in space would therefore be a possible biomarker, indicating its possible production in habitats similar to those found on Earth. The recent discovery of this molecule both in interstellar clouds and in Comet 67P/C-G is interesting and its rejection as a biomarker is significant in the context of what we have earlier (Fayolle, Oberg, Jorgensen, et al., 2017) discussed.

Next, we note the intriguing observation that Comet Lovejoy emits a prodigious amount of ethyl alcohol equivalent to 500 bottles of wine per second (Biver et al., 2015) which has no explanation in terms of simple chemistry. On the other hand many species of fermenting bacteria can produce ethanol from sugars, so this discovery that may well be another indication that microbial processes operate in comets.

Biological activity occurring in comets would be expected to lead to ejecta that reach escape velocities from the entire solar system. The result is that a significant fraction (~1%) of the C, N, O in the solar system that became incorporated in primordial comets (10^{-2}–10^{-3} solar masses) would be returned to space in the form of bacterial dust.

Furthermore, in comets approaching perihelion biological activity could be transiently triggered by solar heating thus leading to the generation of high-pressure pockets of gaseous metabolites that in turn trigger the release of jets of gas and biological particles in their wake. Similar conditions are likely to exist in subsurface "lakes" in other solar system planetary bodies including Europa and Enceladus.

The idea that microbial life cannot exist in comets still dominates scientific orthodoxy but its rational basis is fast eroding. There is ample evidence that microorganisms are resistant to freeze—thaw cycles, to high doses of ionizing radiation, as well to actual space conditions (Hornek, Moeller, Cadet, Douki, et al., 2012). Moreover recent studies have shown that microorganisms can also metabolize in arctic sea ice as well as in permafrost (Junge, Eicken, Swanson, & Deming, 2006).

6. Some predictions of interstellar panspermia

Microorganisms on the Earth display an incredible range of functional diversity. Particular microbial species colonize particular niches, tolerating extremes of temperature, acidity, salinity, desiccation as well as high levels of ionizing radiation. Microbial habitats include subterranean ocean caves, hydrothermal vents, polar ice and tropospheric clouds. Throughout this vast range of conditions an amazing biochemical unity prevails pointing unerringly to a single origin. According to interstellar panspermia that event was external to our solar system.

If life is discovered on planetary habitats outside the Earth such life will be expected to possess the same unity of biochemistry that is found on the Earth—the same genetic code, the same DNA, and identical biochemistries. Only the habitat will select what is best suited for survival, similar to the microorganisms inhabiting the diverse range of habitats on Earth. If independent origination events occur such correspondences will be impossible.

A radically different biochemistry discovered elsewhere could be considered as a falsification of our basic premise.

The new paradigm discussed in this chapter reduces the status of the Earth to that of a trivial building site—one of trillions of similar sites upon which an already evolved cosmological legacy of life came to be expressed. We argue that the most improbable of all events—the origin of life—could not and *did not* happen on the Earth but instead required cosmological vast dimensions of space and time.

As the history of science has repeatedly shown any theory that requires the Earth to hold a special position in the universe has always been proven wrong (the Copernican Principle). Thus the Earth is not the center of our universe, nor is our solar system or our Milky Way galaxy. It is therefore extremely unlikely that the Earth can in any way be regarded the center of life of the universe.

Maintaining the premise of an Earth-centered origin of life is consistent with an outdated Aristotelian world view ("fireflies emerging from a mixture of warm earth and morning dew…") that has held sway for over 2 millennia. The competing idea that life is a cosmic phenomenon is considered inadmissible by many scientists today, but maintaining the status quo appears to demand a multitude of arbitrary fiats, for which there is no hard evidence. We concluding by reminding the reader that a similar situation prevailed before the Ptolemaic solar system was finally abandoned. Historian of science Thomas Kuhn writes thus:

> *"……The state of Ptolemaic (Earth-centered) astronomy was a scandal before Copernicus' announcement. Given a particular discrepancy, astronomers were invariably able to eliminate it by making some particular adjustment in Ptolemy's system of compounded circles. But as time went on, a man looking at the net result of the normal research effort of many astronomers could observe that astronomy's complexity was increasing far more rapidly than its accuracy and that a discrepancy corrected in one place was likely to show up in another…"*
>
> ***Thomas S. Kuhn:* "The Structure of Scientific Revolutions."**

References

Abel, D. L. (2009). The universal plausibility metric (UPM) and principle (UPP). *Theoretical Biology & Medical Modelling*, *6*, 27. https://doi.org/10.1186/1742-4682-6-27.

Abel, D. L., & Trevors, J. T. (2006). Self-organising vs self-ordering events in life-origin models. *Physics of Life Reviews*, *3*(4), 211–228.

Allen, D. A., & Wickramasinghe, D. T. (1981). Diffuse interstellar absorption bands between 2.9 and 4.0 μm. *Nature*, *294*, 239–240.

Altwegg, K., Balsiger, H., Bar-Nun, A., Berthelier, J.-J., Bieler, A., Bochsler, P., et al. (2016). Prebiotic chemical—Amino acid and phosphorus in the coma of comet 67P/Churyumov-Gerasimenko. *Science Advances*, *2*, e1600285.

Bell, E. A., Boehnke, P., Harrison, T., & Mao, W. L. (2015). Potentially biogenic carbon preserved in a 4.1 billion-year-old zircon. *Proceedings of the National Academy of Sciences of the United States of America*, *112*, 14518–14521.

Bieler, K., Altwegg, K., Balsiger, H., Bar-Nun, A., Berthelier, J.-J., Bochsler, P., et al. (2015). Abundant molecular oxygen in the coma of comet 67P/Churyumov-Gerasimenko. *Nature, 526*, 678–681. https://doi.org/10.1038/nature15707.

Biver, N., Bockelee-Morvan, D., Moreno, R., Crovisier, J., Colom, P., Lis, D. C., et al. (2015). Ethyl alcohol and sugar in comet C/2014 Q2 (Lovejoy). *Science Advances, 1*, e150086.

Capaccione, F., Coradini, A., Filacchione, G., Erard, S., Arnold, G., Drossart, P., et al. (2015). The organic-rich surface of comet 67P/Churyumov-Gerasimenko as seen by VIRTIS/Rosetta. *Science, 347*(6220), aaa0628.

Deamer, D. (2011). *First life: Discovering the connections between stars, cells, and how life began.* Berkeley, USA: University of California Press.

Fagerbakke, K. M., Heldal, M., & Norland, S. (1996). Content of carbon, nitrogen, oxygen, sulfur and phosphorus in native aquatic and cultured bacteria. *Aquatic Microbial Ecology, 10*, 15–27.

Fayolle, E. C., Oberg, K. L., Jorgensen, K. K., et al. (2017). Protostellar and cometary detection of organohalogens. *Nature Astronomy, 1*, 703–708.

Hoover, R. B., Hoyle, F., Wickramasinghe, N. C., Hoover, M. I., & Al-Mufti, S. (1986). Diatoms on Earth, comets, Europa and in interstellar space. *Earth, Moon and Planets, 35*, 19–45.

Hornek, G., Moeller, R., Cadet, J., Douki, T., et al. (2012). Resistance of bacterial endospores to outer space for planetary protection purposes—Experiment PROTECT of the EXPOSE-E Mission. *Astrobiology, 12*(5), 445–456.

Hoyle, F. & Wickramasinghe, N.C. (1981). Comets—A vehicle for panspermia, in Comets and the origin of life (ed. C. Ponnamperuma) D. Reidel (reprinted as *Astrophys. Space Sci.,* 268, 333-341 (1999)).

Hoyle, F., & Wickramasinghe, N. C. (1985). *Living comets.* University College Cardiff Press.

Hoyle, F., & Wickramasinghe, N. C. (2000). *Astronomical origins of life: Steps towards panspermia.* Dordrecht: Kluwer Academic Publishers.

Hoyle, F., Wickramasinghe, N. C., & Al-Mufti, S. (1982). Organo siliceous biomolecules in the mid-infrared spectrum of the Trapezium nebula. *Astrophysics and Space Science, 86*, 63–69.

Hoyle, F., & Wickramasinghe, N. C. (1991). *The theory of cosmic grains.* Dordrecht: Kluwer Academic Press.

Junge, K., Eicken, H., Swanson, B. D., & Deming, J. W. (2006). Bacterial incorporation of leucine into protein down to −20°C with evidence for potential activity in sub-eutectic saline. *Cryobiology, 52*, 417–429.

Laturnus, F. (2001). Marine macroalgae in polar regions as natural sources for volatile organohalogens. *Environmental Science and Pollution Research, 8*(2), 103–108.

Laturnus, F., Adaams, F. C., & Wiencke, C. (1998). Methylhalides from Antarctic macroalgae. *Geophysical Research Letters, 25*, 773–776.

Lobert, J. M., Keene, W. C., Logan, J. A., & Yevich, R. (1999). Global chlorine emissions from biomass burning: Reactive chlorine emissions inventory. *Journal of Geophysical Research, 104*, 8373–8389.

Miller, S. L., & Urey, H. C. (1959). Organic compound synthesis on the primitive earth. *Science, 130*, 245–251.

Motta, V., Mediavilla, E., Munoz, J. A., Falco, E., Kochanek, C. S., Arribas, S., et al. (2002). Detection of the 2175A extinction feature at z=0.83. *The Astrophysical Journal, 574*, 719–725.

Noterdaeme, P., Petitjean, P., Ledoux, C., & Srianand, R. (2009). Evolution of the cosmological mass density of neutral gas from Sloan Digital Sky Survey II—Data release 7. *Astronomy and Astrophysics, 505*, 1087–1098.

Okuda, H., Shibai, H., Nakagawa, T., Matsuhara, H., Kobayashi, Y., Kaifu, N., et al. (1990). An infrared quintuplet near the galactic center. *The Astrophysical Journal, 351*, 89–97.

Rauf, K., & Wickramasinghe, C. (2009). Evidence for biodegradation products in the interstellar medium. *International Journal of Astrobiology*, *9*, 29–34.
Steele, E. J., Al-Mufti, S., Augustyn, K. A., Chandrajith, R., Coghlan, J. P., Coulson, S. G., et al. (2018). Cause of Cambrian explosion—terrestrial or cosmic? *Progress in Biophysics and Molecular Biology*, *136*, 3–23.
Steele, E. J., Gorczynski, R. M., Lindley, R. A., Liu, Y., Temple, R., Tokoro, G., et al. (2019). Lamarck and panspermia—On the efficient spread of living systems throughout the cosmos. *Progress in Biophysics and Molecular Biology*, *149*, 10–32.
Teplitz, H. I., Desai, V., Armus, L., Chary, R., Marshall, J. A., Colbert, J. W., et al. (2007). Measuring PAH emission in ultradeep spitzer IRS spectroscopy of high redshift IR luminous galaxies. *The Astrophysical Journal*, *659*, 941–949.
Wallis, M. K., & Wickramasinghe, N. C. (2015). Pluto's surprises: Mountain tectonics, methane, and evidence of biology. *Journal of Astrobiology and Outreach*, *3*, 3. https://doi.org/10.4172/2332-2519.1000135.
Watling, R., & Harper, D. B. (1998). Chloromethane production by wood-rotting fungi and an estimate of the global flux to the atmosphere. *Mycological Research*, *102*, 769–787.
Wickramasinghe, N. C. (1967). Interstellar Grains *(Chapman & Hall, London)*.
Wickramasinghe, N. C. (1974). Formaldehyde polymers in interstellar space. *Nature*, *252*, 462–463.
Wickramasinghe, C. (2011). Viva panspermia. *The Observatory*, *131*, 130–134.
Wickramasinghe, D. T., & Allen, D. A. (1986). Discovery of organic grains in Comet Halley. *Nature*, *323*, 44–46.
Wickramasinghe, D. T., Hoyle, F., Wickramasinghe, N. C., & Al-Mufti, S. (1986). A model of the 2-4 micron spectrum of Comet Halley. *Earth, Moon, Planets*, *36*, 295–299.
Wickramasinghe, N. C., & Tokoro, G. (2014). Life as a cosmic phenomenon: 1 The socio-economic control of a scientific paradigm. *Journal of Astrobiology and Outreach*, *2*, 133.
Wickramasinghe, N. C., Wainwright, M., Smith, W. E., et al. (2015). Rosetta studies of Comet 67P/Churyumov–Gerasimenko: Prospects for establishing cometary biology. *Journal of Astrobiology and Outreach*, *3*, 126.

Further reading

Crovisier, J., Leech, K., Bockelee-Morvan, D., Brooke, T. Y., Hanner, M. S., Altieri, B., et al. (1997). The spectrum of Comet Hale-Bopp (C/1995 01) observed with the Infrared Space Observatory at 2.9 AU from the Sun. *Science*, *275*, 1904–1907.
Kuhn, T. (1970). *The structure of scientific revolutions*. University of Chicago Press.

CHAPTER THREE

The efficient Lamarckian spread of life in the cosmos

Edward J. Steele[a,b,c,*], Reginald M. Gorczynski[d], Robyn A. Lindley[e,f], Yongsheng Liu[g], Robert Temple[h], Gensuke Tokoro[l], Dayal T. Wickramasinghe[j], and N. Chandra Wickramasinghe[b,i,k,m,*]

[a]C.Y.O'Connor ERADE Village Foundation, Piara Waters, Perth, WA, Australia
[b]Centre for Astrobiology, University of Ruhuna, Matara, Sri Lanka
[c]Melville Analytics Pty Ltd, Melbourne, VIC, Australia
[d]Department of Surgery & Immunology, University of Toronto, Toronto, ON, Canada
[e]Department of Clinical Pathology, Faculty of Medicine, Dentistry & Health Sciences, University of Melbourne, Melbourne, VIC, Australia
[f]GMDx Group Ltd, Melbourne, VIC, Australia
[g]Henan Collaborative Innovation Center of Modern Biological Breeding, Henan Institute of Science and Technology, Xinxiang, China
[h]The History of Chinese Science and Culture Foundation, Conway Hall, London, United Kingdom
[i]Institute for the Study of Panspermia and Astroeconomics, Gifu, Japan
[j]College of Physical and Mathematical Sciences, Australian National University, Canberra, ACT, Australia
[k]University of Buckingham, Buckingham, United Kingdom
[l]Institute for the Study of Panspermia and Astroeconomics, Gifu, Japan
[m]National Institute of Fundamental Studies, Kandy, Sri Lanka
*Corresponding authors: e-mail address: e.j.steele@bigpond.com; ncwick@gmail.com

Contents

Abstract

In this Chapter we discuss the various mechanisms that are available for the possible transfer of cosmic microbial living systems from one cosmic habitat to another. With the 100 or so habitable planets that are now known to exist in our galaxy alone transfers of cometary dust carrying life including fragments of icy planetoids/asteroids would be expected to occur on a routine basis. It is thus easy to view the galaxy as a single connected "biosphere" of which our planet Earth is a minor component.

Advances in Genetics, Volume 106
ISSN 0065-2660
https://doi.org/10.1016/bs.adgen.2020.03.004

The Hoyle–Wickramasinghe Panspermia paradigm provides a cogent biological rationale for the actual widespread existence of Lamarckian modes of inheritance in terrestrial systems (which we review here). Thus the Panspermia paradigm provides the *raison d'etre* for Lamarckian Inheritance. Under a terrestrially confined neoDarwinian viewpoint such an association may have been thought spurious in the past. Our aim here is to outline the main evidence for rapid terrestrial-based Lamarckian-based evolutionary hypermutation processes dependent on reverse transcription-coupled mechanisms among others. Such rapid adaptation mechanisms would be consistent with the effective cosmic spread of living systems. For example, a viable, or cryo-preserved, living system traveling through space in a protective matrix will of necessity need to *adapt rapidly* and proliferate on landing in a new cosmic niche. Lamarckian mechanisms thus come to the fore and supersede the slow (blind and random) genetic processes expected under neoDarwinian Earth centred theories.

1. The traditional neoDarwinian view of life and evolution

It is useful then to begin by summarizing the widely held traditional view of the evolution of life on Earth. In this way we can sharpen the focus on the set of Lamarckian acquired inheritance processes now discovered which, together with Panspermic "in-falls" of living systems from space, are now replacing the neoDarwinian explanation for the origin life and its further evolution on Earth. A dot point summary would go something like this:

- Life emerged in the period after the Hadean Epoch (about 4 billion years ago) as the first free-living cells, arising from non-living chemistry (Abiogenesis), proliferated on Earth.
- This key Abiogenic event happened 3–4 billion years ago perhaps via an RNA World, in one of Darwin's hypothetical "warm little ponds" or the "primordial soup."
- These free-living primordial cells then proliferated and flourished by Darwinian evolution. They repeatedly duplicated their genomes and genes, with additional slow mutational events occurring. Survival of the fittest of these duplicated derivatives was the origin of new DNA sequences and the rearrangements occurring in their genomes, new species.
- The progression is envisaged to be from the archaeal and prokaryotic bacterial worlds, generating many new cellular species. The production of viruses by most cells went hand in glove with this evolution and further aided cell-cell genetic communication. Thus virus mediated

horizontal gene transfer (HGT) was a key early process aiding cellular diversification and adaptation.

- Darwinian natural selection is seen therefore as preserving all these evolutionary genetic steps which arose by chance and random events. We do not doubt that at least some of these sorts of slow genetic events may have happened.
- The first free living eukaryotic cells occurred by cell-cell fusions between archaeal and prokaryotic cells by symbiosis (Margulis, 1970, 1981). It is then envisaged that these events provided the evolutionary path prior to the explosive emergence of the metazoans and multicellular plant and animal life in the Cambrian period ca 540 million years ago.
- With the apparent exception of the Cambrian explosion itself, however, many of these evolutionary phases envisaged were very slow, taking millions if not hundreds of millions of years.
- Note that the Cambrian "creative disruption" to this linear thinking is not an exception, just a big one. The *actual fossil record*, as well as phylogenetic nucleic acid and protein sequence analyses, show that most novel species and life forms emerge suddenly in a "punctuated" way, either persisting to the present or going to extinction (termed "punctuated equilibrium" by Eldridge and Gould, published in 1972, 1977; the same disrupted events are discussed in Hoyle & Wickramasinghe, 1981, and again in Steele et al., 2018).
- Indeed ever since the explosive events of the Cambrian adaptive radiation two further major extinctions and adaptive radiations are well preserved in the fossil record, and occurred at the Permian/Triassic (P/T) boundary (~252 million years ago) and the Cretaceous/Palaeocene (K/T) boundary (~65.5 million years ago, with the extinction of Dinosaurs).
- Such apocalyptic events seem to have a frequency of occurring approximately every 250 hundred million years, suggesting a cosmic orbiting cycle of our Sun and Solar System around the galactic center of the Milky Way. There are also shorter 30 million year extinction-diversification cycles suggesting an alignment as our star system oscillates through the galactic plane (Clube, Hoyle, Napier, & Wickramasinghe, 1996; Wickramasinghe et al., 2010).

But we need to ask ourselves, is this a credible way to view the evolution of life on Earth? Abiogenesis on Earth, or anywhere in the known observable Universe is highly improbable (the odds against are less than 1 successful event expected in $>10^{5120}$ events, Hoyle & Wickramasinghe, 1999, and outlined in Steele, Al-Mufti, et al., 2019, Steele, Gorczynski, et al., 2019).

An alternative view, which is argued below, is one of scientific pragmatism rather than ideological pursuit of improbable events or just so scenarios. Our key evidence for Lamarckian inheritance, which overcomes the pitfalls inherent to theories of a slow and ponderous neoDarwinian evolutionary progression for the origin of life on Earth, is outlined in Tables 1 and 2.

Our purpose here then is to show that since the 1970s many key lines of scientific evidence have emerged contradicting this comfortable view of life on Earth. We and many colleagues have recently reviewed most of this salient contradictory evidence (Steele et al., 2018; Steele, Al-Mufti, et al., 2019; Steele, Gorczynski, et al., 2019). We conclude that the traditional terrestrial neoDarwinian view of terrestrial evolution is scientifically untenable. At key junctures it does not fit the actually observed data showing that "...living organisms such as space-resistant and space-hardy bacteria, viruses, more complex eukaryotic cells, fertilized ova and seeds have been continuously delivered ... to Earth so being one important driver of further terrestrial evolution which has resulted in considerable genetic diversity and which has led to the emergence of mankind" (Steele et al., 2018). Thus the evolution of life on Earth has been driven by seedings from the wider Cosmos. These space-derived "varieties" also further evolve on Earth and drive further diversification. Every few hundreds of millions of years on Earth is marked by major Cosmic bolide "seeding" events presumably as mentioned by passage of our Solar System through giant dust and molecular clouds (e.g., see Hoyle & Wickramasinghe, 1993).

In our view the key hard evidence already exists forming the "Demarcation Data" distinguishing terrestrial neo-Darwinism from the evidence for Cosmic Panspermia. This evidence from the 1970s to 2000 has been marshalled and analyzed by Hoyle and Wickramasinghe (1978, 1979, 1981, 1985, 1993, 1999 which references all key peer-reviewed papers). Wickramasinghe (2018) has reviewed all the key evidence covering the terrestrial atmosphere, bacteria and other micro-organisms, discussing a role for comets as protective incubators and amplifiers of living systems, and highlighted among the evidence the infra-red (IR) analyses of interstellar dust and cometary ejecta among other central issues.

We have also highlighted some important 'Demarcation Data' that distinguishes neo-Darwinism from Cosmic Panspermia (Steele, Al-Mufti, et al., 2019; Steele, Gorczynski, et al., 2019). One key focus is on the internal structure of carbonaceous meteorites showing clear microfossils of both eukaryotic and prokaryotic organisms. These observations have been from independently curated and examined carbonaceous meteorites, dated at

Table 1 Adaptive genetic strategies of environmentally-driven mutators systems (main Ref. Steele, Gorczynski, et al., 2019).

System	Adaptive genetic process	Molecular, enzymatic, cellular processes involved in adaptation	References
Immune system	• Innate immunity (Germline encoded)—to intracellular pathogens (e.g., viruses) and neo-antigens generated in cancer cells by somatic mutation	• AID/APOBEC C-to-U and ADAR A-to-I (G) deaminases mutate pathogen/host genomes. These single nucleotide variants (SNVs) are lesions which generate base pair mismatches which if left unrepaired lead to single nucleotide changes. If the repair process is incomplete it can lead to single stand DNA nicks which if unrepaired lead to chromosome rearrangements and DNA translocations.	Refsland and Harris (2013), Samuel (2011), Schoggins and Rice (2011), Schneider, Chevillotte, and Rice (2014), Lindley (2013), Lindley and Steele (2013, 2018), Lindley, Humbert, Larmer, Akmeemana, and Pendlebury (2016)
	• Acquired adaptive immunity-somatic hypermutation (SHM) of rearranged immunoglobulin (Ig) V regions and, in principle rearranged T cell receptor (TCR) V regions (Termed "VDJs") allowing specific responses to foreign antigens on pathogens (viruses, bacteria) or novel host antigens (neo-antigens) on cancer cells	• AID/APOBEC C-to-U and ADAR A-to-I (G) deaminases initiate and somatically mutate VDJs. Dysregulated Ig SHM like responses typical of cancer genomes. • Can include target site reverse transcription (TSRT) via Y family DNA polymerases, e.g., Pol-η, Pol-κ. • Homologous recombination (HR)	Steele (2016a), Steele and Lindley (2017), Franklin, Steele, and Lindley (2020)

Continued

Table 1 Adaptive genetic strategies of environmentally-driven mutators systems (main Ref. Steele, Gorczynski, et al., 2019).—cont'd

System	Adaptive genetic process	Molecular, enzymatic, cellular processes involved in adaptation	References
	• Origin, maintenance, diversification of germline rearranged V gene repertoires	• AID/APOBEC C-to-U and ADAR A-to-I (G) deaminases initiate and somatically mutate VDJs. • Can include target site reverse transcription (TSRT) via Y family DNA polymerases, e.g., Pol-η, Pol-κ. • Soma-to-germline flow DNA/RNA sequences via vesicles, endogenous retroviral vectors • Homologous recombination (HR)	Steele and Lindley (2018), Steele and Lindley (2020)
Human Genome	• Origin of neutral, benign, beneficial germline variants including those causing human genetic diseases viz. origin of single nucleotide polymorphisms (SNPs) in OMIM and dbSNP databases	• AID/APOBEC C-to-U and ADAR A-to-I (G) deaminases (direct and/or indirect effect germline DNA sequences (protein coding in particular). Elements of other processes listed above viz. Can include Y family polymerases Pol-η, Pol-κ, reverse transcription also (TSRT), and homologous recombination (HR)	Lindley and Hall (2018)

Table 2 Adaptive genetic strategies of environmentally-driven mutators systems (main Ref. Steele, Gorczynski, et al., 2019).

System	Adaptive genetic process	Molecular, enzymatic, cellular processes involved in adaptation	References
Vertebrate immune and endocrine systems	Inheritance acquired tolerance, immunity and behavior	• Acquired transgenerational inheritance • Penetration Weismann barrier • Sire effect-coupled maternal effect • Pavlovian conditioning-couple to acquired inheritance	Gorczynski and Steele (1980, 1981), Gorczynski et al. (1983), Sobey and Conolly (1986), Gorczynski (1992)
	Inheritance of acquired instincts "Dutch Famine"	• Acquired epigenetic inheritance • Acquired epigenetic-genetic effects	Campbell and Perkins (1988), Jablonka & Lamb, 1995), Lindley (2010), Painter et al. (2008), Pembrey, Saffery, and Bygren (2014), Dias and Ressler (2014)
	Inheritance acquired autoimmunity	• Transmission via male of maternal induced autoimmune eye defects (Experiments of Guyer & Smith 1918–1924)	Reprinted in Steele 2016b
Mammalian germ cells	Uptake of foreign DNA and RNA	• Penetration Weismann barrier • Soma-to-Germline flow DNA/RNA with genetic effects on progeny	Spadafora (1998, 2008, 2018), Rassoulzadegan et al. (2006), Fogarty (2002)
Plant hybrid grafts	Acquired inheritance	• Classic pangenesis in hybrid plants	Liu (2018)
Adaptive evolution in bacteria and microorganism	Adaptive hypermutation SOS responses	• Substrate, ligand, stress-induced hypermutation, rapid "Darwinian" selection. Putative involvement Y family DNA polymerases, thus Reverse Transcription (?)	Radman (1974, 1999), Tippin, Phuong Pham, and Goodman (2004), Rosenberg (2001)

>4.5 billion years old. The data have been independently confirmed by experts in four well curated and characterized carbonaceous meteorites: Murchison (Hoover, 2005, 2011; Pflug & Heinz, 1997), Murchison, Orgueil, Mighei (Rozanov & Hoover, 2013), Polonnaruwa (Wallis et al., 2013; Wickramasinghe, Wallis, Wallis, & Samaranayake, 2013). We stress all this, as the scepticism often raised here actually threatens a key corner stone of Science viz. the independent confirmation of evidence—without this criterion modern Science could not function nor progress.

In all these meteorite studies terrestrial contamination has been ruled out. So we must seriously consider these four different and independent "experiments of nature" on their own terms. Eukaryotic fossils with silica-based hard shells, are prominent in these fossils. In the Polonnaruwa meteorite (Wickramasinghe et al., 2013) the frustules of clear diatoms are evident (Fig. 1). Clearly these are features of mature cell biology in astrophysical phenomena that require a coherent explanation. "They strongly imply that complex cell-based life, now immortalized as fossils in carbonaceous meteorites, pre-dates the age of the Earth (and solar system). An explanation based on Panspermia seems unavoidable to us" (Steele, Al-Mufti, et al., 2019; Steele, Gorczynski, et al., 2019).

We have argued pragmatically (Steele, Gorczynski, et al., 2019) that the origins from non-living chemistry of cell-based life in the current known universe are so statistically improbable that our scientific approach should

Fig. 1 Eukaryotic microfossil in Carbonaceous meterorite. Voidal-shaped ribbed structure embedded in the rock matrix of the Polonnaruwa carbonaceous meteorite, Wickramasinghe et al. (2013), see also other chemical analyses in Wallis et al. (2013).

Table 3 From Steele, Gorczynski, et al. (2019)—Projected numbers based on Earth equivalents and estimated number of stars in the observable Universe $\sim 10^{22}$.

Cosmic distribution and numbers of living systems	
• Viruses—terrestrial number 10^{31}	10^{53}
• Bacteria/Archaea—terrestrial number $> 10^{30}$	10^{52}
• Single cell eukaryotes—terrestrial number 10^{20}–10^{30}	10^{32}–10^{52}
• Complex Metazoans—terrestrial number $\geq 10^{20}$	10^{42}
• Higher plants, terrestrial number $\geq 10^{7}$ species	10^{29}
• Higher animals, terrestrial number $\geq 10^{7}$ species	10^{29}

not be directed so much toward studying abiogenesis events in the laboratory, but into quantitating the distribution and numbers of living systems in the observable Universe (e.g., Table 3, and Steele, Gorczynski, et al., 2019). Indeed there is much to learn pragmatically about life in the Cosmos by observing what falls to Earth from space (see chapter "Origin of new emergent Coronavirus and Candida fungal diseases—Terrestrial or cosmic?" by Steele et al., this volume), along with other biological evidence for living systems in the near Earth neighborhood, e.g., bacteria in cosmic dust on the external surface of the International Space Station (Grebennikova et al., 2018).

2. The rise of the neoLamarckian acquired inheritance paradigm

We have critically reviewed much of the evidence for Lamarckian inheritance gathered since the 1970s which have established the new Lamarckian evolutionary paradigm (Steele, Gorczynski, et al., 2019).

We have condensed the main lines of this evidence in Tables 1 and 2. These environmentally-induced cellular and molecular processes can now be considered the *primary evolutionary driver mechanisms* for the evolution and ongoing diversification of life on Earth. And, by extension, we argue the case here (see chapter "The sociology of science and generality of the DNA/RNA/protein paradigm throughout the cosmos" by Wickramasinghe et al., this volume) and elsewhere (Wickramasinghe, Wickramasinghe, Tout, Lattanzio, & Steele, 2020) that these DNA and RNA inheritance mechanisms are likely to be general throughout the Cosmos—resulting in the efficient Panspermic dispersal of living systems, in all their variety, throughout the Universe.

Table 4 From Steele, Gorczynski, et al. (2019).

Evidence consistent with Lamarckian evolutionary processes
1. Environmental stimulation as the directional mutational driver
2. Role of epigenetic gene targeting
3. Rapid genetic adaptation
4. Penetration of the *Weismann Barrier*
5. Horizontal gene transfer (HGT)
6. Central role of reverse transcription

The summaries of evidence for horizontal gene transfer (HGT) are covered at the Wikipedia site https://en.wikipedia.org/wiki/Horizontal_gene_ transfer.

The key distinguishing features and driver mechanisms of this neoLamarckian view of life are listed in Table 4.

We now describe pertinent molecular-cellular evidence for Lamarckian acquired-inheritance adaptations across the kingdoms of life—from bacterial cells to higher plants and animals. The review will not be comprehensive because of the vastness of the literature since the 1970s. We will select representative examples for discussion and illumination as outlined in Tables 1 and 2.

2.1 Immune system

2.1.1 Innate immunity

This is the first line of defense against invading pathogens in all cells and is thus termed "Innate" immunity. As the name implies the functions of Innate Immunity are germline selected and encoded and thus not "adaptive" somatically over time, but immediately "reactive" to foreign pathogen entry to cells. Probably a 100 or so genes are co-ordinately upregulated during an Innate Immune response via interferon mediated cytokines, and they are termed Interferon Stimulated Genes (ISGs, Schneider et al., 2014). Key ISG components of Innate Immunity are the cytosine deaminases (AID/APOBEC) converting cytosine in DNA to uracil (C-to-U) and adenosine (ADAR) deaminases converting adenosines to inosines in double stranded (ds)RNA, or in RNA and DNA moieties of RNA:DNA hybrids that occur in Transcription Bubbles (Steele & Lindley, 2017; Zheng, Lorenzo, & Beal, 2017). The inosine bases so formed are mutagenic (A-to-I, read as "G" on replication, reverse transcription or during translation from mRNA to protein sequences, see chapter "The mutagenic source and power of our own evolution" by Lindley, this volume). Thus both mutagenic uracils and inosines need to be repaired prior to DNA or RNA copying events- when this does not happen it causes G•T, G•U and A•I base pair mismatches in the DNA and RNA helices (leading to C-to-T and A-to-G primary mutations). Downstream DNA repair sequelae of these lesions to single stranded DNA nicks (or staggered double stranded nicks) leads to DNA rearrangements and translocation events (common in progressing cancer genomes). Further, the error prone translesion Y family DNA Polymerases eta (η) and kappa (κ) in mammalian systems are also efficient reverse transcriptases (Franklin, Milburn, Blanden, & Steele, 2004; Franklin et al., 2020) so that error prone reverse transcriptase Y family copying of such lesions can lead to transversion mutations, e.g., A-to-T, A-to-C.

Invading viral genomes replicating in the cytosolic endoplasmic reticulum membranes (so called Replication Factories) or similar membranous sites within the nucleus are the main targets for deamination during the Innate Immune phase (first 24–48 h after infection). The cytosine and adenosine deaminases thus cause mutagenic damage to pathogen genomes. The C-to-U and A-to-I(G) mutational signatures at the appropriate deaminase sequence motifs are literally written all over the viral genomes in the early phase of infection (e.g., see Lindley & Steele, 2018). It is now thought that these aggressive deaminase responses can also cause host genome damage leading to somatic mutations in host cell genes and facilitating cancerous growth on further progression (Lindley, 2013; Lindley & Steele, 2013, Lindley, Humbert et al., 2016, and see chapter "The mutagenic source and power of our own evolution" by Lindley, this volume). It also seems that the components of the Adaptive Immune Response (somatic hypermutation, SHM) are co-opted, if not highjacked, as "Dysregulated Ig - SHM -like responses", now cause off-target mutations across the genome in genes expressed in cells far beyond the lymphoid system.

2.1.2 Acquired immunity

The variable (V) genes of higher vertebrate antibodies, or immunoglobulins (Ig), encode the antigen-binding regions of antibody (and T cell receptor) proteins. They exist as large arrays in the germline DNA. These numerous very similar V sequences (≥50–100 V gene segments) in the germline are normally inactive. However when they are rearranged at the DNA level in a mature somatic B (or T) lymphocyte in the lymphoid and blood circulation they become transcriptionally active. The Vs join with shorter D and J elements forming a transcriptionally active somatic gene encoding rearranged heavy (VDJ) and light (VJ) chains of the HL heterodimers of antibody proteins. There are many possible DNA rearrangement combination in the clonal progenitor cell of a B or T lymphocyte lineage: thus for antibodies, with say 100 germline Vs, 10 germline Ds, and 5 germline Js in a heavy chain there are 100 × 10 × 5 or 5000 possible heavy chain encoding VDJs with clean joints (no joining errors). For a light chain there would be maybe 100 Vs, 5–10 Js so this will give about 500–1000 VJs (light chains). A viable antigen combining site is formed from a protein heterodimer of one heavy (H) and light (L) chains, essentially by a combinatorial protein association and sorting process in any given B cell. So the potential starting somatic repertoire of antibody specificities is large of the order of 10^6 antibody clonal specificities available for selection by invading antigens early in a response. However such V[D]J genes are also stimulated into somatic hypermutation (SHM), particularly in B lymphocytes (but see Steele & Lindley, 2020).

It is in this "somatic" configuration that V[D]J genes hypermutate following antigenic stimulation. This is a typical "Darwinian" process of rapid mutation, proliferation of antigen-selected B cell survivors with large cellular apoptotic death factors (in so called "Germinal Centres" in peripheral lymphoid organs, such as spleen, lymph nodes).

Thus many mutated B cells are destined to die (>90%) in Germinal Centres. The successful antigen-selected mutants, bearing an antigen-specific receptor on their surface membrane survive to become affinity-improved, clonally expanded, memory B lymphocytes (clonal selection). The daughter cells then enter the vascular circulation and seed other lymphoid organs. All the extant in vivo molecular and cellular evidence indicates the mechanism of Ig somatic hypermutation (SHM) is driven by antigenic stimulation via an AID (APOBEC) and ADAR deaminase-coupled target site reverse transcription process (TSRT), as listed in Table 1.

2.1.3 Germline V repertoires

The origin, maintenance and further diversification of these mammalian germline V arrays, for both immunoglobulins (Ig, the antibodies made by B lymphocytes) and the receptors on T cells (TCRs) which recognize and kill viral/pathogen infected host target cells is a major unsolved problem in immunology and genetics. It now appears that these germ line V gene arrays evolve via a reverse transcriptase coupled soma-to-germline feedback process (Steele & Lindley, 2018, 2020). This we predict can be detected by the active real-time transgenerational penetration of the Weismann Barrier in extended families (Steele & Lloyd, 2015). Thus, all the elements of the deaminase-mediated mutation process and TSRT via Y family DNA polymerases appear involved at some level. The final genetic insertion steps into the germline DNA would, by necessity, need to be via a targeted homologous recombination process (HR). This step is discussed in the cited references (Table 1, and in Steele, Gorczynski, et al., 2019) and particularly in Steele & Lloyd, 2015).

2.2 Human genome

The origin of the numerous (millions) single nucleotide variations (polymorphisms or SNPs) between the germlines of different human individuals is a major issue in biology, now revealed in this age of rapid and cheap whole genome next generation sequencing. Many of the SNPs in protein-coding regions are functionally neutral, but many are functionally subtle and benign, some are overtly beneficial (and termed "positively Darwinian selected"). These germline variants include those causing human genetic diseases viz. origin of single nucleotide polymorphisms (SNPs) curated in the Online Mendelian Inheritance in Man db (OMIM) and the NCBI dbSNP human genome database (National Center for Biotechnology Information, Bethesda, Maryland, USA). When critically analyzed a vast portion of these SNPs can be accounted for by deamination events at the major known cytosine (AID, APOBEC3G, APOBEC3B) and adenosine (ADAR) deamination motifs (Lindley & Hall, 2018). The important implication is either that AID/APOBEC and ADAR deaminases act in a profligate manner, always active in the transcribed genes of human germline cells, or that these single nucleotide variants originate first in somatic cells and are selected somatically for functional integration into the organism's physiology: that is, AID/APOBEC C-to-U and ADAR A-to-I (G) deaminases are active on expressed protein-coding genes in somatic cells and the successful mutant sequences are then transferred to the germline. Elements of other molecular processes listed appear also to be involved since a RNA-to-DNA reverse transcription step would need to occur for many of the A-site ADAR deamination events (viz. Y family polymerases Pol-h, Pol-k, reverse transcription also (TSRT), and homologous recombination HR).

2.3 Vertebrate immune and endocrine systems

2.3.1 Inheritance acquired tolerance, immunity and behavior

Experiments have demonstrated the inheritance of acquired neonatal tolerance to foreign histocompatibility antigens in mice. Deep neonatal induced tolerance of immune reactivity at the level of cytotoxic T lymphocytes

(CTL) could be induced in neonatal male mice which on mating to females of the same inbred stain, passed on their specific immune tolerance to first and second generation progeny (appearing in the second generation without exposure to the foreign histocompatibility (H-2) antigens used to set up specific tolerance in the original father (Gorczynski & Steele, 1980, 1981). In later experiments the specific H-2 tolerance in progeny generations correlated with specific delayed skin graft rejection (Gorczynski, Kennedy, MacRae, & Ciampi, 1983). Later experiments showed positive paternal transmission was demonstrated in experiments in inbred mice to foreign (rat) erythrocytes, using both repeated high dose neonatal male tolerance (Steele, Gorczynski, & Pollard, 1984) or single shot immunity to the erythrocytes in adult males prior to breeding to normal females (Steele, 1984). These complex transgenerational effects almost certainly involve both genetic and epigenetic regulation resulting in hard genetic inheritance dimensions (as discussed in Steele, Gorczynski, et al., 2019, fig. 2 in that paper).

Acquired inheritance phenomena associated with what is called the "Sire Effect" have been reviewed (Lindley, 2010 pp. 22–29) and the phenomenon was first reported by Gorczynski et al. (1983) when they tested the normal inbred female mice who had raised offspring to male mice of the same inbred strain made neonatally tolerant to repeated doses of foreign lymphoid cells expressing specific H-2 histocompatibility antigens as just described (Gorczynski & Steele, 1980, 1981). When these mothers when bred to normal males of the same inbred strain they produced tolerant or hyporesponsive progeny to the *same H-2 antigens* as used in the original neonatal tolerance regime in the original breeding male.

These mothers also passed on the effect when fostering normal pups, identifying causal factors in the colostrum and milk (Gorczynski et al., 1983). This striking result has implications resurrecting the old observations surrounding the non-Mendelian breeding results caused by male sperm and thus phenomena associated with "Telegony" (Watson, Carroll, & Chaykin, 1983). The "Sire Effect" thus opens a Pandora's Box with wide implications for pure-line animal breeding and wider societal implications (Lindley, 2010 pp. 22).

The "Sire Effect "has real-world practical implications (Lindley, 2010 pp. 26–27). Wild rabbits in rural Australia were in plague proportions in the 1940s and 1950s causing great damage to agriculture, particularly the sheep and cattle industries. To control these wild rabbit populations the Commonwealth Scientific and Industrial Research Organization (CSIRO) released rabbits infected with lethal *Myxomatosis* virus. The virus was very effective initially in controlling wild rabbit numbers. But the speed with which immunological resistance developed caused further investigations. At first sight it seemed much faster than simple "Darwinian" recovery of a resistant residual population after such a large kill ($\geq$90%). Indeed careful follow up controlled breeding experimental work by Bill Sobey and Dorothy Connolly discovered a significant factor in the rapid spread of resistance: bucks which had recovered from *Myxomatosis* virus when mated to a doe who had not previously been exposed to *Myxomatosis* virus produced litters that were resistant to the lethal effects of the virus- a clear paternal transmission or "Sire Effect" as described by Gorczynski et al. (1983). Thus when a non-immune buck was mated with a doe that had previously been mated to an immune buck a significant number of progeny were born with immunity to *Myxomatosis* virus. Sobey and Conolly concluded that an unknown factor transmitted via the semen of the *Myxomatosis* virus recovered bucks to the normal females which could be further transferred in other matings to normal non-exposed males.

The molecular-cellular mechanisms of the *Myxomatosis* virus sire effect in rabbits, as well as the H-2 antigen-specific maternal influence in the mother's milk acquired by the mother from the original neonatally tolerant male have not been analyzed. Obvious candidates for study can be drawn from an array of epigenetic and genetic transmission effects discussed in Steele, Gorczynski, et al. (2019), fig. 2. They could be related to the small regulatory RNA-mediated spermatozoa effects described by Rassoulzadegan and colleagues (Kiani et al., 2013; Liebers, Rassoulzadegan, & Lyko, 2014; Rassoulzadegan et al., 2006) and other foreign RNA and DNA in semen and associated with spermatozoa reported by Spadafora (Spadafora, 1998, 2008, 2018).

2.3.2 Inheritance acquired instincts

How do instincts arise in evolution? Logic implies an ultimate adaptive Lamarckian cause in ancestors. Strong survival instincts in parents based on prior learnt fear responses must have arisen in our ancestors not by random chance events that were specifically selected, but in a Lamarckian manner, which were then passed on to their progeny *en masse* providing a survival value to the small familial and interbreeding groups of mammals in the wild. Dias and Ressler (2014) showed that parental mice subjected to Pavlovian odor fear conditioning before conception produced offspring with specific behavioral sensitivity to the specific chemical odor used to condition the parents. Unrelated chemical odors did not trigger a conditioned fear response, and these specific acquired transgenerational effects are indeed inherited via parental gametes. It appears both direct genomic and indirect epigenetic odorant receptor gene targeting appear to act together ("epigenetic-genetic" coupling) to establish what we now recognize as specific instinctual responses involving odorant receptor genes and behavior.

2.3.3 Dutch famine

This well-known consequences of starvation in Holland at the end of World War II are a well documented human epigenetic transgenerational phenomenon, on scale. These extreme starvation episodes underscore the long terms epigenetic inherited effects (Painter et al., 2008). We would also predict that longer term hard inheritance genetic effects are buried in this human data (Steele, Gorczynski, et al., 2019).

2.3.4 Inheritance acquired autoimmunity

A hundred years ago Guyer and Smith at the University of Wisconsin published their classic experiments showing hereditary transmission via male rabbits of antibody induced eye defects though the mother down nine breeding generations viz. maternal induced autoimmune eye defects. A summary of these experimental results by Guyer and Smith are republished in detail with a commentary in Steele (2016b). We are of the view that in the fullness of time these experiments will join the garden pea experiments of Mendel and the plant graft hybrid experiments of Charles Darwin as classics in genetics, particularly acquired inheritance associated with cosmic genetic evolution.

2.4 Mammalian germ cells

2.4.1 Uptake of foreign DNA by spermatozoa and inherited effects in progeny

Corrado Spadafora and colleagues have shown sperm uptake of foreign nucleic acid molecules and transmission of the genetic information to progeny organisms. In some cases they show that a LINE-1-derived reverse transcription step can execute the copying the RNA into DNA. In $\leq$10% of cases the DNA sequences are integrated into the germline genome, while in most cases the sperm-absorbed DNA/RNA exists as extrachromosomal episomes (replicating along with host somatic cells during development and displaying mosaic tissue expression). A role was postulated for exosomes vesicles released into the bloodstream from human tumor xenografts transferring somatic RNA to spermatozoa, implying that there is no physical barrier in spermatozoa to the uptake of DNA or RNA. This work is among the most precise descriptions of Darwin's "gemmules" in animals published, and is a mode of vesicle transfer which is also utilized in genetic information transfer in plant graft-hybridization, described below (Liu, 2018) and consistent with the current way we now view Lamarckian inheritance. Finally in 2002 Patrick Fogarty reported a striking result of simple intraperitoneal injection of DNA into adult male or female mice. Employing a technique based on P-element transposons and delivering DNA transgenes intravenously in simple vesicles, Fogarty has shown that 50% of progeny from such male mice inherit the gene sequence (Fogarty, 2002). The critical integration event requires a transposase enzyme. This was a clear demonstration of the genetic penetration of the *Weismann Barrier* in mammals in a Lamarckian mode typical of a familiar Horizontal Gene Transfer event.

2.5 Graft-induced heritable changes in plants

2.5.1 Acquired inheritance

Historically graft-induced variations were recorded to occur in ancient China. Charles Darwin however was the first to use the term "graft hybridization". He noted that the formation of breeding hybrids through plant grafting between distinct species or varieties (without the intervention of the sexual organs). Liu (2018) outlines all of these observations that have been confirmed now many times by experts. The results are clear and striking and regularly reproduced by many plant breeders and horticulturists. Indeed, plant hybrid transmission of small regulatory RNAs, mRNAs and other reporter sequences are standard experimental tools in plant molecular genetics and development (Ham & Lucas, 2017). Many such cases of "graft hybrids", were described by Darwin where shoots produced from grafted plants exhibited a combination of characters of both rootstock and scion. It was understandable that he would explain their formation by his theory of Pangenesis involving transmissible and transported "gemmules" in the phloem. Over the past few decades, the existence of graft hybrids has been widely documented, and graft hybridization is considered to be a simple and efficient means of plant breeding. Graft hybridization is now explained by horizontal gene transfer and DNA transformation and the long-distance transport of mRNA and small RNAs considered to be involved in the formation of graft hybrid seeds. We should now add a reverse transcription step to lock in such transported RNA phloem information into the hybrid seed genomes.

2.6 Adaptive evolution in bacteria and microorganisms

2.6.1 Adaptive hypermutation and SOS genetic responses

Developments separate from the above, are less definitive conceptually, involving the penetration of the *Weismann Barrier*, is the work in bacteria and other rapidly multiplying unicellular organisms (e.g., yeast *spp*). In these cases simple Darwinian selection of rare population variants can never be ruled out. However very challenging demonstrations of substrate-induced adaptive evolutionary phenomena in bacteria and yeast have been described (Cairns, Overbaugh, & Miller, 1988; Hall, 1988; Rosenberg, 2001). The data suggest rapid mutator mechanisms in microorganims during the stationary phase (slower replication) which increases the odds of 'selecting' an adaptive mutant. The data have been interpreted as being consistent with a Lamarckian process of adaptive evolutionary genetic change and the authors of the studies suggest a reverse transcriptase-coupled mechanism for the inheritance of acquired characteristics, much like that proposed earlier by Steele and Pollard (1987) for the somatic hypermutation processes summarized in Steele, Gorczynski, et al. (2019) and Table 1.

This work raised the possibility that environmentally induced, nonrandom mutator processes dependent on a reverse transcriptase step also occur in bacteria and yeast. For bàcterial spp. this has been confirmed and many different reverse transcriptase activities have now been described throughout prokaryotes and archaea, reviewed in Steele, Gorczynski, et al. (2019). Indeed Radman (1999) had earlier predicted such error prone polymerase enzymes of evolutionary change (Radman, 1974) as part of the now familiar adaptive "SOS response" in bacteria to a range of environmental stress signals (Tippin et al., 2004). It is intriguing that the Y family DNA Polymerases that figure prominently in the SOS response are all error-prone polymerases and are indeed related by their DNA encoded sequence to the human Y family DNA repair polymerases eta (η), kappa (κ), iota (ι) (Ohmori et al., 2001) all of which have been shown to be efficient reverse transcriptases (Franklin et al., 2004). However none of the other bacterial Y family DNA polymerase members have yet to be examined for their RT activity.

3. Conclusions

We have summarized here some of the new evidence that has accumulated since the 1970s consistent with Lamarckian Acquired Inheritance phenomena from bacteria through to plants and animals. We are of the view that the "Hypothesis of Lamarck" has now survived some stringent and severe tests earning recognition as a maturing "Theory" of biological evolution. The results of current studies on the molecular mechanisms of these phenomena are variable, and in different stages of advancement. But the mechanisms, and the validity of transgenerational epigenetic-genetic coupled phenomena are, despite our ignorance, much clearer now in many cases than they were 40 years ago. And our discussion and analyses of the relevant extant data in Steele, Gorczynski, et al. (2019) in the context of Cosmic Panspermia, outlined above, clearly needs to be confronted by the scientific mainstream. Lamarckian inheritance of acquired characteristics, which at its core depends on base modified RNA intermediates and reverse transcriptase steps, has successfully run the gauntlet of numerous "Popperian" tests during the past 40 years. We argue then that the effective spread of living systems throughout the Cosmos is both by Lamarckian and Darwinian mechanisms—with the emphasis on Lamarckian evolutionary processes as these provide rapid and "directional" adaptations to new Cosmic niches immediately after the organisms have landed.

References

Cairns, J., Overbaugh, J., & Miller, S. (1988). The origin of mutants. *Nature, 335*, 142–145.

Campbell, J. H., & Perkins, P. (1988). Transgenerational effects of drug and hormonal treatments in mammals: A review of observations and ideas. In: Boer, G.J., Feenstra, M.G.P., Mirmiran, M., Swaab, D.F., van Haaren, F. (Eds.) *Progress in Brain Research, 73*, 535–553.

Clube, S. V. M., Hoyle, F., Napier, W. M., & Wickramasinghe, N. C. (1996). Giant comets, evolution and civilization. *Astrophysics and Space Science, 245*, 43–60.

Dias, B. G., & Ressler, K. J. (2014). Parental olfactory experience influences behavior and neural structure in subsequent generations. *Nature Neuroscience, 17*, 89–96.

Eldredge, N., & Gould, S. J. (1972). Punctuated equilibria: An alternative to phyletic gradualism. In T. J. M. Schopf (Ed.), *Models in paleobiology* (pp. 82–115). San Francisco: Freeman Cooper.

Fogarty, P. (2002). Optimizing the production of animal models for target and lead validation. *Targets, 1*(3), 109–116. https://doi.org/10.1016/S1477-3627(02)02198-0.

Franklin, A., Milburn, P. J., Blanden, R. V., & Steele, E. J. (2004). Human DNA polymerase-h an A-T mutator in somatic hypermutation of rearranged immunoglobulin genes, is a reverse transcriptase. *Immunology and Cell Biology, 82*, 219–225.

Franklin, A., Steele, E. J., & Lindley, R. A. (2020). A proposed reverse transcription mechanism for (CAG)n and similar expandable repeats that cause neurological and other diseases. *Heliyon, 6*, e03258. https://doi.org/10.1016/j.heliyon.2020.e03258.

Gorczynski, R. M. (1992). Conditioned stress responses by pregnant and/or lactating mice reduce immune responses of their offspring after weaning. *Brain, Behavior, and Immunity, 6*, 87–95.

Gorczynski, R. M., Kennedy, M., MacRae, S., & Ciampi, A. (1983). A possible maternal effect in the abnormal hyporesponsiveness to specific alloantigens in offspring born to neonatally tolerant fathers. *Journal of Immunology, 131*, 1115–1120.

Gorczynski, R. M., & Steele, E. J. (1980). Inheritance of acquired immunologic tolerance to foreign histocompatibility antigens in mice. *Proceedings of the National Academy of Sciences of the United States of America*, *77*, 2871–2875.
Gorczynski, R. M., & Steele, E. J. (1981). Simultaneous yet independent inheritance of somatically acquired tolerance to two distinct H-2 antigenic haplotype determinants in mice. *Nature*, *289*, 678–681. https://doi.org/10.1038/289678a0.
Gould, S. J., & Eldredge, N. (1977). Punctuated equilibria: The tempo and mode of evolution reconsidered. *Paleobiology*, *3*, 115–151.
Grebennikova, T. V., Syroeshkin, A. V., Shubralova, E. V., Eliseeva, O. V., Kostina, L. V., et al. (2018). The DNA of bacteria of the world ocean and the Earth in cosmic dust at the international space station. *The Scientific World Journal2018*. https://doi.org/10.1155/2018/7360147, Article ID 7360147, 7 pp.
Hall, G. H. (1988). Adaptive evolution that requires multiple spontaneous mutations. 1. Mutations involving an insertion sequence. *Genetics*, *120*, 887–897.
Ham, B.-K., & Lucas, W. J. (2017). Phloem-mobile RNAs as systemic signaling agents. *Annual Review of Plant Biology*, *68*, 173–195.
Hoover, R. B. (2005). Microfossils, biominerals, and chemical biomarkers in meteorites. In R. B. Hoover, A. Y. Rozanov, & Paepe (Eds.), *Perspectives in astrobiology* (pp. 43–65). Amsterdam: RR IOS Press.
Hoover, R. B. (2011). Fossils of cyanobacteria in CI1 carbonaceous meteorites: impli-cations to life on comets, Europa and Enceladus. *Journal of Cosmology*, *16*, 7070–7111.
Hoyle, F., & Wickramasinghe, N. C. (1978). *Life cloud*. London: J.M. Dent Ltd.
Hoyle, F., & Wickramasinghe, N. C. (1979). *Diseases from space*. London: J.M. Dent Ltd.
Hoyle, F., & Wickramasinghe, N. C. (1981). *Evolution from space*. London: J.M. Dent Ltd.
Hoyle, F., & Wickramasinghe, N. C. (1985). *Living comets*. Cardiff: Univ. College, Cardiff Press.
Hoyle, F., & Wickramasinghe, N. C. (1993). *Our place in the cosmos : The unfinished revolution*. London: J.M. Dent Ltd.
Hoyle, F., & Wickramasinghe, N. C. (1999). Panspermia 2000. *Astrophysics and Space Science*, *268*, 1–17. In Hoyle, F., & Wickramasinghe, N.C. (2000). Astronomical Origins of Life: Steps towards Panspermia. Panspermia 2000. Preface to Reprints from Astrophys. Space Sci 268. Kluwer Academic Publishers, Dordrecht, The Netherlands, pp. 1-3, 1999.
Jablonka, E., & Lamb, M. J. (1995). *Epigenetic inheritance and evolution: The Lamarckian dimension*. Oxford: Oxford University Press.
Kiani, J., Grandjean, V., Liebers, R., Tuorto, F., Ghanbarian, H., et al. (2013). RNA-mediated epigenetic heredity requires the cytosine methyltransferase Dnmt2. *PLoS Genetics*, *9*(5), e1003498. https://doi.org/10.1371/journal.pgen.1003498.
Liebers, R., Rassoulzadegan, M., & Lyko, F. (2014). Epigenetic regulation by heritable RNA. *PLoS Genetics*, *10*, e1004296. https://doi.org/10.1371/journal.pgen.1004296.
Lindley, R. (2010). *The Soma: How our genes really work and how that changes everything!*. CYO Foundation. POD book. CreateSpace, Amazon.com. ISBN 1451525648.
Lindley, R. A. (2013). The importance of codon context for understanding the Ig-like somatic hypermutation strand-biased patterns in TP53 mutations in breast cancer. *Cancer Genetics*, *206*, 222–226. https://doi.org/10.1016/j.cancergen.2013.05.016.
Lindley, R. A., & Hall, N. E. (2018). APOBEC and ADAR deaminases may cause many single nucleotide polymorphisms curated in the OMIM data- base. *Mutation Research*, *810*, 33–38.
Lindley, R. A., Humbert, P., Larmer, C., Akmeemana, E. H., & Pendlebury, C. R. R. (2016). Association between targeted somatic mutation (TSM) signatures and HGS-OvCa progression. *Cancer Medicine*, *5*, 2629–2640. https://doi.org/10.1002/cam4.825.
Lindley, R. A., & Steele, E. J. (2013). Critical analysis of strand-biased somatic mutation signatures in TP53 versus Ig genes, in genome-wide data and the etiology of cancer. *ISRN*

Genomics, 2013 Article ID 921418, 18 pages. https://www.hindawi.com/journals/isrn/2013/921418/.
Lindley, R. A., & Steele, E. J. (2018). ADAR and APOBEC editing signatures in viral RNA during acute- phase innate immune responses of the host-parasite relationship to Flaviviruses. *Research Reports, 2*, e1–e22. https://doi.org/10.9777/rr.2018.10325. The Supp data is at the journal site http://www.companyofscientists.com/index.php/rr.
Liu, Y. (2018). Darwin's pangenesis and graft hybridization. *Advances in Genetics, 102*, 27–66.
Margulis, L. (1970). *Origin of eukaryotic cells*. New Haven, CT: Yale University Press, ISBN 0–300–01353-1.
Margulis, L. (1981). *Symbiosis in cell evolution*. San Francisco: W.H. Freeman and Co.
Ohmori, H., Friedberg, E. C., Fuchs, R. P. P., Goodman, M. F., Hanaoka, F., Hinkle, D., et al. (2001). The Y-family of DNA polymerases. *Molecular Cell, 8*, 7–8.
Painter, R., Osmond, C., Gluckman, P., Hanson, M., Phillips, D., & Roseboom, T. (2008). Transgenerational effects of prenatal exposure to the Dutch famine on neonatal adiposity and health in later life. *BJOG, 115*, 1243–1249.
Pembrey, M., Saffery, R., & Bygren, L. O. (2014). Human transgenerational responses to early-life experience: Potential impact on development, health and biomedical research. *Journal of Medical Genetics, 51*, 563–572.
Pflug, H. D., & Heinz, B. (1997). Analysis of fossil organic nanostructures: Terrestrial and extraterrestrial. *SPIE Proceedings Instruments, Methods, and Missions for the Investigation of Extraterrestrial Microorganisms, 86*(3111), 1997. https://doi.org/10.1117/12.278814. July11.
Radman, M. (1974). In L. Prakashet al. (Ed.)*Molecular and environmental aspects of mutagenesis* (pp. 128–142). Springfield, Illinois: C. C. Thomas.
Radman, M. (1999). Enzymes of evolutionary change. *Nature, 401*, 866–869.
Rassoulzadegan, M., Grandjean, V., Gounon, P., Vincent, S., Gillot, I., et al. (2006). RNA-mediated non-mendelian inheritance of an epigenetic change in the mouse. *Nature, 441*, 469–474.
Refsland, E. W., & Harris, R. S. (2013). The APOBEC3 family of retroelement restriction factors. *Current Topics in Microbiology and Immunology, 371*, 1–27. https://doi.org/10.1007/978-3-642-37765-5_1.
Rosenberg, S. M. (2001). Evolving responsively: adaptive mutation. *Nature Reviews. Genetics, 2*, 805–815.
Rozanov, A. Y., & Hoover, R. B. (2013). Acritarchs in carbonaceous meteorites and terrestrial rocks. Instruments, methods, and missions for astrobiology XVI. In R. B. Hoover, G. V. Levin, A. Y. Rozanov, & N. C. Wickramasinghe (Eds.), *Proceedings of SPIE.*, Vol. 8855. https://doi.org/10.1117/12.2029608, 886507-7.
Samuel, C. E. (2011). Adenosine deaminases acting on RNA (ADARs) are both antiviral and proviral. *Virology, 411*, 180–193. https://doi.org/10.1016/j.virol.2010.12.004.
Schneider, W. M., Chevillotte, M. D., & Rice, C. M. (2014). Interferon-stimulated genes: A complex web of host defenses. *Annual Review of Immunology, 232*(2014), 513–545. https://doi.org/10.1146/annurev-immunol-032713-120231.
Schoggins, J. W., & Rice, C. M. (2011). Interferon-stimulated genes and their antiviral effector functions. *Current Opinion in Virology, 1*(6), 519–525. https://doi.org/10.1016/j.coviro.2011.10.008.
Sobey, W. R., & Conolly, D. (1986). Myxomatosis; nongenetic aspects of resistance to myxomatosis in rabbits, Oryctogagus cuniculus. *Australian Wildlife Research, 13*, 177–187.
Spadafora, C. (1998). Sperm cells and foreign DNA: A controversial relation. *BioEssays, 20*, 955–964.
Spadafora, C. (2008). Sperm-mediated "reverse" gene transfer: A role of reverse transcriptase in the generation of new genetic information. *Human Reproduction, 23*, 735–740. https://doi.org/10.1093/humrep/dem425.

Spadafora, C. (2018). The "evolutionary field" hypothesis. Non-Mendelian trans- generational inheritance mediates diversification and evolution. *Progress in Biophysics and Molecular Biology, 134*, 27–37. https://doi.org/10.1016/j.pbiomolbio.2017.12.001.

Steele, E. J. (1984). Acquired paternal influence in mice. Altered serum antibody response in the progeny population of immunized CBA/H males. *Immunology and Cell Biology, 62*, 253–268.

Steele, E. J. (2016a). Somatic hypermutation in immunity and cancer: Critical analysis of strand-biased and codon-context mutation signatures. *DNA Repair, 45*, 1–24. https://doi.org/10.1016/j.dnarep.2016.07.001.

Steele, E. J. (2016b). Origin of congenital defects: Stable inheritance through the male line via maternal antibodies specific for eye lens antigens inducing autoimmune eye defects in developing rabbits *in utero*. In M. Levin & D. S. Adams (Eds.), *Ahead of the curve-hidden breakthroughs in the biosciences*. Bristol, UK: Michael Levin and Dany Spencer Adams IOP Publishing Ltd, Chapter 3.

Steele, E. J., Al-Mufti, S., Augustyn, K. A., Chandrajith, R., Coghlan, J. P., Coulson, S. G., et al. (2018). Cause of Cambrian explosion - terrestrial or cosmic? *Progress in Biophysics and Molecular Biology, 136*, 3–23.

Steele, E. J., Al-Mufti, S., Augustyn, K. A., Chandrajith, R., Coghlan, J. P., Coulson, S. G., et al. (2019). Cause of Cambrian explosion - terrestrial or cosmic?—Reply to commentary by R Duggleby. *Progress in Biophysics and Molecular Biology, 141*, 74–78.

Steele, E. J., Gorczynski, R. M., Lindley, R. A., Liu, Y., Temple, R., Tokoro, G., et al. (2019). Lamarck and Panspermia—On the efficient spread of living systems throughout the Cosmos. *Progress in Biophysics and Molecular Biology, 149*, 10–32.

Steele, E. J., Gorczynski, R. M., & Pollard, J. W. (1984). The somatic selection of acquired characters. In J. W. Pollard (Ed.), *Evolutionary theory: Paths into the future* (pp. 217–237). London: John Wiley.

Steele, E. J., & Lindley, R. A. (2017). ADAR deaminase A-to-I editing of DNA and RNA moieties of RNA:DNA hybrids has implications for the mechanism of Ig somatic hypermutation. *DNA Repair, 55*, 1–6.

Steele, E. J., & Lindley, R. A. (2018). Germline V repertoires: Origin, maintenance, diversification. *Scandinavian Journal of Immunology, 87*, e12670. https://doi.org/10.1111/sji.12670.

Steele, E. J., & Lindley, R. A. (2020). Regulatory T cells and co-evolution of allele-specific MHC recognition by the TCR. *Scandinavian Journal of Immunology, 91*, e12853. https://doi.org/10.1111/sji.12853.

Steele, E. J., & Lloyd, S. S. (2015). Soma-to-germline feedback is implied by the extreme polymorphism at IGHV relative to MHC. *BioEssays, 37*, 557–569.

Steele, E. J., & Pollard, J. W. (1987). Hypothesis: Somatic hypermutation by gene con- version via the error-prone DNA->RNA->DNA information loop. *Molecular Immunology, 24*, 667–673.

Tippin, B., Phuong Pham, P., & Goodman, M. F. (2004). Error-prone replication for better or worse. *Trends in Microbiology, 12*, 2988e2995. https://doi.org/10.1016/j.tim.2004.04.004.

Wallis, J., Miyake, N., Hoover, R. B., Oldroyd, A., Wallis, D. H., Samaranayake, A., et al. (2013). The Polonnaruwa meteorite; Oxygen isotope, crystalline and biological composition. *Journal of Cosmology, 22*(2) published, 5 March 2013.

Watson, J. G., Carroll, J., & Chaykin, S. (1983). Reproduction in mice: The fate of sperm cells not involved in fertilization. *Gamete Research*, 7, 75–84.

Wickramasinghe, C. (2018). *Proofs that Life is Cosmic.* Singapore: World Scientific Publishing Co.

Wickramasinghe, N. C., Wallis, J., Wallis, D. H., & Samaranayake, A. (2013). Fossil diatoms in a new carbonaceous meteorite. *Journal of Cosmology, 21*(37) published, 10 January 2013.

Wickramasinghe, J. T., Wickramasinghe, N. C., & Napier, W. M. (2010). *Comets and the and the origin of life*. Singapore: World Scientific Publishing Co.
Wickramasinghe, N. C., Wickramasinghe, D. T., Tout, C. A., Lattanzio, J. C., & Steele, E. J. (2020). Cosmic biology in perspective. *Astrophysics and Space Science,* In Press. https://arxiv.org/abs/1805.10126.
Zheng, Y. C., Lorenzo, C., & Beal, P. A. (2017). DNA editing in DNA/RNA hybrids by adenosine deaminases that act on RNA. *Nucleic Acids Research, 45*, 3369–3377.

Further reading

Hoyle, F., & Wickramasinghe, C. (1982). *Why Neo-Darwinism does not work*. University College Cardiff Press, ISBN 0 906449 50 2.
Hoyle, F., & Wickramasinghe, N. C. (1991). *The theory of cosmic grains*. Dordrecht: Kluwer.

CHAPTER FOUR

The sociology of science and generality of the DNA/RNA/protein paradigm throughout the cosmos

N. Chandra Wickramasinghe[a,b,c,d,*], Edward J. Steele[e,f,*], Brig Klyce[g], Gensuke Tokoro[b], and Dayal T. Wickramasinghe[h]

[a]Centre for Astrobiology, University of Ruhuna, Matara, Sri Lanka
[b]Institute for the Study of Panspermia and Astroeconomics, Gifu, Japan
[c]University of Buckingham, Buckingham, United Kingdom
[d]National Institute of Fundamental Studies, Kandy, Sri Lanka
[e]C.Y.O'Connor ERADE Village Foundation, Piara Waters, Perth, WA, Australia
[f]Melville Analytics Pty Ltd, Melbourne, VIC, Australia
[g]Astrobiology Research Trust, Memphis, TN, United States
[h]College of Physical and Mathematical Sciences, Australian National University, Canberra, ACT, Australia
*Corresponding authors: e-mail address: ncwick@gmail.com; e.j.steele@bigpond.com

Contents

Abstract

The theory of cometary panspermia argues that life cannot have originated on Earth in the time available. It must have an ultimate, but still undiscovered cosmological source. The origin of life remains an open question. Life on Earth was introduced by impacting comets, and its further evolution was driven by the subsequent acquisition of cosmically derived genes. Explicit predictions of this theory stating how the acquisition of new genes drives evolution, are compared with recent developments in relation to horizontal gene transfer, and the role of retroviruses in evolution. Precisely stated predictions of the theory of cometary panspermia are shown to have been verified.

Advances in Genetics, Volume 106
ISSN 0065-2660
https://doi.org/10.1016/bs.adgen.2020.03.005

1. The socio-economic control of science

"Falsehood and delusion are allowed in no case whatever: but, as in the exercise of all the virtues, there is an economy of truth. It is a sort of temperance, by which a man speaks truth with measure that he may speak it the longer...."

Burke (1849)

Economy of Truth is a principle of limitation often used by politicians whenever the Whole Truth is deemed strategically unwise. We have demonstrated by implication throughout this volume that the same principle is often used in 21st century science as a mode of controlling the flow of information, and the mechanism of control involves the collective and often covert decisions of large and diffuse societal groups. The neo-Darwinian evolutionary paradigm is a case in point. It was dominant for over a century but in recent years it has suffered a substantial decline and collapse, as the theory does not fit the observed facts nor is it able to predict how life on Earth actually started and evolved (Chapter 3, Steele et al., 2019).

In modern times the involvement of the state or of large organizations in the conduct of science has become necessary to varying degrees. This is due mainly to the requirement of funds to set up laboratories, which are often expensive and beyond the reach of individual scientists. Moreover, the so-called "big" projects in all areas of science require large teams using expensive equipment. So, organization and central control becomes imperative. Examples of ongoing big projects include the space exploration of planets by NASA and other similar space agencies, the Hadron Collider operated by CERN, and major genome sequencing projects in several countries—and recently the control of infectious diseases as far as this is possible.

In its earliest beginnings science arose as the solitary pursuit of individual philosophers whose ideas were often opposed to the status quo. The pre-Socratic philosopher Anaxagoras in the 5th century BCE declared that the Sun was a red-hot stone and the Moon was made of earth, and for his heresy he was banished from Athens. State-control of science is therefore no new thing. Examples are to be found scattered throughout history—extending from the time of classical Greece, through the long saga of the Ptolemaic epicycles in the Middle Ages, the control of science by the Papacy, and to some degree, stretching through into modern times. Although academic freedom is enshrined in most university statutes today, its implementation often leaves much to be desired.

There are many aspects of the conduct of 21st century science that are similar to the behavior of a totalitarian state. A totalitarian regime in politics sets out a rigid framework of rules to govern society, and a system of law for punishing those who disobeyed. Transgressions being met with severe penalties implied that there was always a firm motive for citizens to conform. Communist regimes, such as exist in the Soviet Union and China, fit well into this general pattern. Marxist-Leninist philosophy was based on denouncing the perceived evils of capitalism, and its implementation involved a great deal of state control of property and resources, as well as, unfortunately, severe restrictions in freedom of speech.

While in the spheres of politics and economics such state control may have a justification as a prerequisite for firm government, a similar control extending to other areas of creativity including art, music and science, is less desirable, and may act in a way that impedes progress and stifles innovation. The justification of eugenics in Nazi Germany with its grotesque and inhumane consequences and the enforcement of obscurantist biological theories in the Soviet Union, provide examples of such conduct. Biology under Marxism also vigorously defended the principle of spontaneous generation despite the fact that this principle was essentially disproved by the experimental work of Louis Pasteur in the 1860s.

On the other hand the Marxist-Russian promulgation of "Lysenkoism" was in many ways correct as environmentally driven Lamarckian inheritance is now a wide spread phenomenon that has been vindicated in animals, plants and even microorganisms display aspects (adaptive evolution) of the rapid adaptation phenomenon (Chapter 3, Steele et al., 2019). And this is particularly so with the wide spread recognition of genuine transgenerational inheritance and induced epigenetic phenomena in plants and animals (Liu, 2018; Steele et al., 2019).

Ideas of the Russian biologist Alexander Oparin that led to the theory of the origin of life in a primordial soup were undoubtedly inspired by the tenets of dialectical materialism. Oparin and the Soviet scientists drew their inspiration from the German philosopher Friedrich Engels who had proposed that new qualities of "being" arose at each new stage of organic evolution. Engels noted that higher levels of existence resulted from lower levels, and this progression was deemed part of the natural order of things. The primordial soup paradigm of the origin of life derived from this philosophy still remains the reigning dogma in science although its political and philosophical antecedents are now largely forgotten.

2. Transition to modern times

Modern science has taken on a totally different complexion, where progress depends crucially on expensive equipment, large teams of workers, and the support, direct or indirect, of large organizations sponsored by the state. If ideas ran counter to those of an influential majority, or a powerful establishment, progress will be severely hampered. This is true both in a capitalist system as well as under communist dictatorships—summed up in the phrase "every country has its own Gulag." In either case the control of new ideas is what one would expect within a totalitarian political system. Dissent from a majority position in science is quickly and effectively quelled by starvation of funds, chastisement of those attempting to promote contrary views, and more invidiously by preventing unpopular ideas from becoming widely known.

If all this is true, how, one might ask, is scientific progress still taking place, seemingly at an astounding pace? To answer this question, it is useful to divide science into several types. The type of empirical/predictive science that informs us how matter—living or non-living—behaves is the kind of science that we routinely learn at school and university. The mechanics of Newton, atomic and nuclear physics, the well-attested properties of matter and radiation, do not offer themselves as subjects of political dispute of any kind. It is upon this kind of science that the entire structure of modern technology depends. Although biology at a molecular level—e.g., DNA, sequencing—comes in the same category, the bigger organizational and inferential structures of biology—e.g., theories of the origin and evolution of life, as well as of epidemic diseases—lend themselves to manipulation, by political and scientific authorities.

3. Cosmic biology as a scientific theory

A scientific theory has value only if it can make testable predictions. Falsifiability, according to fulfillment or otherwise of predictions, is the essence of the process by which a convergence to an objective truth can be achieved. The cometary panspermia hypothesis and the overarching theory of comic biology developed from the 1980s onwards challenged the earlier prevailing dogma of an origin of life in a "warm little pond" on Earth. The super-astronomical complexity of life at a molecular genetic level prompts one to question this hypothesis on probabilistic grounds (see *Foreword* by Klyce, and a most recent review of pragmatic arguments in Steele et al., 2019). Indeed, the universality of the Genetic Code on Earth, convinced Crick and Orgel (1973) that the DNA/RNA/protein paradigm had to be a cosmic universal. They advocated a form of intelligent 'Directed Panspermia' for seeding life on Earth.

So the theory of the cosmic origins of life based on the uniformity of the DNA/RNA/protein paradigm made a prediction that the infrared and ultraviolet spectra of interstellar dust and comets would match biologically derived material (the detritus of life). This prediction was verified convincingly over several decades of study and the fact that it is still not regarded as proof of the theory relates to the sociological issues we discussed earlier.

The predictions of the cosmic life theory that will be reviewed in the present chapter are those related to the evolution of terrestrial life that were originally discussed by Hoyle and Wickramasinghe (1979, 1981). It was argued that if comets brought the first life to Earth 4 billion years ago, the process of microbial additions from comets must have continued throughout geological time, and consequently played a role in the evolution of life. Such considerations were later extended to include a model where genetic products of evolution on a planet like the Earth were mixed on a galactic scale with products of local evolution on other distant, extrasolar planets.

It is possible to argue that comet impacts, such as happened at the K/T boundary 65 million years ago leading to the extinction of the dinosaurs, leads also to the inevitable splash back into space of DNA fragments that carry the products of local evolution (Napier, 2004; Wallis & Wickramasinghe, 2004; Wickramasinghe, Wickramasinghe, & Wallis, 2010). Even partially inactivated bacterial and viral genes expelled in this way may be capable of transmitting useful genetic information to recipient organisms on a distant planet (Wesson, 2010; Wickramasinghe, 2011). We could argue that similar impact episodes on other inhabited planets would be inevitable and so the products of biological evolution would be mixed on a galactic scale. In such a scheme, we would inevitably have a uniformity of DNA/RNA/protein based biology established on a galactic scale.

We know that our solar system orbits the centre of the galaxy with a period of 240 million years. The average time interval between successive encounters of the solar system with massive dust clouds is about 40 million years, so that during the lifetime of the solar system (4500 million years) over a hundred such gene distribution events would have taken place for the Earth alone. Since we cannot assume the Earth and our solar system to unique in this regard, it has to be assumed that similar gene dissemination sources exist for every life-bearing Earth-like planet in the galaxy. As a consequence the biosphere in which Darwinian/Lamarckian evolution occurs is naturally extended to encompass a large fraction of the volume of the galactic disc. The stochastic nature of gene acquisition events resulting from encounters with molecular clouds in the galaxy leads naturally to a stochastic component in biological evolution—e.g., sudden jumps, as is apparently observed on the Earth, the Cambrian explosion being one such example.

A firm prediction of this theory was that extraterrestrial viruses and bacteria—occasionally responsible for pandemics of disease—became incorporated in the germ line of survivors and provided the main driving force for evolution (Hoyle & Wickramasinghe, 1979, 1990). Although Hoyle and Chandra Wickramasinghe were ridiculed at the time for apparently advocating a return to a primitive superstition with regard to comets and plagues, and for challenging geocentric ideas about life, advances in biology in the new millennium have produced striking evidence in favor of this point of view. Of particular importance is the discovery of horizontal gene transfer operating across a wide range of phyla, as well as of genome sequencing including the sequencing of the human genome. To begin with, we shall recap in Section 4, certain key predictions of the cosmic life theory with direct quotations from publications dating back to 1979 and 1982.

4. Predictions from 1979 and 1982

In *Diseases from Space* (1979) pp. 153, 154 Hoyle and Wickramasinghe wrote:

> *"There are four steps in the complex process whereby such a virus multiplies itself the preventing of any one of which would confer immunity on an evolved plant or animal. The virus must first have an attachment protein fitted to an attachment site on the wall of a host cell. Second, the interaction of the attachment protein to the cell wall must serve eventually to strip other viral proteins away from the genetic material of the virus, which must then be afforded naked ingress to the host cell. Third, the viral genetic material must have the ability to overwrite the normal genetic program of the cell. And fourth, after multiplying in number, the new virus particles must be able to gain egress from the host cell in order to attack new cells."*
>
> *"With so many opportunities to frustrate the attack of viruses, and yet with evolved life forms failing to avail themselves convincingly of these opportunities, we have - within the Earth-bound point of view - the makings of a contradiction. Immunity, such as it is, consists mostly of preventing the entry of virus particles, and then only the entry of specific viruses, not of viruses in general. It would be far more effective*

for host cells to develop genetically so as to prevent overwriting by viruses in general. Logically, such a process must exist, because the greater quantity of information present in the genetic material of the host must be able to overcome the combined information of many viruses. Yet such a defence is never presented, except possibly very transiently during the course of an infection, when a substance called 'interferon' may have some such effect. The more likely explanation is that host cells deliberately avoid the apparently best form of defence. The situation is not at all that the virus is clever but that the host appears to be incorrigibly stupid. Indeed host cells even seem to invite the invasion of viruses by deliberately providing sites to which viruses can attach themselves..."

In *Proofs that Life is Cosmic*, pp. 73, 74 Hoyle and Wickramasinghe (1982) wrote:

"Some commentators have claimed that pathogenic viruses cannot be incident from space, for an imagined reason which they believe overrides the many facts which prove otherwise. The argument seems on minimal thought to have the attractive quality of a one-line disproof. Viruses are specific to the cells they attack it is said, as if to claim that human viruses are specific to human cells. While a minority of human viruses might be said to be specific to the cells of primates, most human viruses can actually be replicated in tissue cell cultures taken from a wide spectrum of animals, some indeed outside the mammals entirely. The proper statement therefore is that viruses are generally specific to the cells they attack to within about 150 million years of evolutionary history ..."

"If we had knowledge that evolution was an entirely terrestrial affair then of course it would be hard to see how viruses from outside the Earth could interact in an intimate way with terrestrially-evolved calls, but we have no such knowledge, and in the absence of knowledge all one can say is that viruses and evolution must go together. If viruses are incident from space then evolution must also be driven from space. How can this happen? Viruses do not always attack the cells they enter. Instead of taking over the genetic apparatus of the cell in order to replicate themselves, a viral particle may add itself placidly to one or other of the chromosomes. If this should happen for the sex cells of a species, mating between similarly infected individuals leads to a new genotype in their offspring, since the genes derived from the virus are copied together with the other genes whenever there is cell division during the growth of the offspring ..."

"A gene that happens to be useful to the adaptation of one life-form may be useless to another. Incidence from space knows nothing of such a difference, however, the gene being as likely to be added to the one form as the other. So genes that become functional in some species may exist only as nonsense genes in other species. This again is true. Genes that are useful to some species are found as redundant genes in other species. Suppose a new gene or genes to become added to the genotype (genome) of a number of members of some species. Suppose also that one or more of the genes could yield a protein or proteins that would be helpful to the adaptation of the species. The cells of those members of the species possessing the favourable new genes operate, however, in accordance with the previously existing genes, a problem arises as to how the new genes are to be switched into

operation so as to become helpful to the species.....As potentially favourable genes pile up more and more, a species acquires a growing potential for large advantageous change, it acquires the potential for a major evolutionary leap, thereby punctuating its otherwise continuing state of little change – it's 'equilibrium' ..."

5. Horizontal gene transfer (HGT)

While material expelled from comets could supply living cells (bacteria) and viruses to a habitable planet like the Earth, the genetic components of evolved life on planets could also be disseminated on a galaxy-wide scale. The entire solar system, which at present is surrounded by an extended shell consisting of some 100 billion comets (the Oort Cloud), orbits around the center of the galaxy with a period of approximately 240My. On the average, every 40 million years, this cloud of comets becomes gravitationally perturbed due to the close passage of a massive molecular cloud such as for instance the Orion nebula. Gravitational interaction leads to hundreds of comets being injected from the cometary shell into the region of the inner planets, some of which would inevitably collide with the Earth. These collisions can not only lead to the extinctions of species (as one such impact surely did 65 million years ago, killing the dinosaurs), but they may also result in the expulsion of surface material laden with the detritus of life as well as terrestrial microbiota of various kinds.

A fraction of the Earth-debris so expelled that survives shock-heating and could be laden with viable microbial ecologies as well as genes (DNA, RNA, proteins) of evolved life. Such life-bearing terrestrial material could reach newly forming planetary systems in the passing molecular cloud within a million years of the ejection event. Neighboring planetary system thus can be infected with terrestrial microbes and terrestrial genes that can contribute, via the process of horizontal gene transfer, to an ongoing process of local biological evolution. Once life has got started and evolved on an alien planet or planets, the same process can be repeated (again via comet collisions) transferring genetic material carrying local evolutionary "experience" to other molecular clouds and other nascent planetary systems. If every life-bearing planet transfers genes in this way to more than one other planetary system (say 1.1 on the average) with a characteristic time of 40My then the number of seeded planets after 9 billion years (lifetime of the Galaxy) is $(1.1)^{9000/40} \sim 2 \times 10^9$. Such a vast number of 'infected' planets illustrates that Darwinian/Lamarckian evolution, involving horizontal gene transfers, must operate not only on the Earth or even within the limited confines of the solar system but on a truly galactic scale. Life throughout the galaxy on this picture would constitute a single connected biosphere (Hoyle & Wickramasinghe, 1979, 1982).

The cosmic theory of life will therefore be inevitably based on an all-pervasive DNA/RNA/protein biology which implies that genes which are the products of evolution in a distant cosmic location can, on occasion, be transferred to evolving lifeforms on another planet such as the Earth (Hoyle & Wickramasinghe, 1979, 1982). In this way, evolutionary advantage or novelty could be acquired by terrestrial organisms on a stochastic basis whenever alien genetic material carrying new information is introduced to the Earth. This leads naturally a galaxy-wide process of gene transfers—transfer of genetic information across normal mating barriers on a cosmological scale. This mechanism was proposed for the burgeoning field of astrobiology in 1982, even before it was demonstrated to operate as a process within terrestrial biology (Hoyle & Wickramasinghe, 1982).

The mechanism of horizontal gene transfer (HGT) has now been amply documented on a terrestrial scale (Keeling & Palmer, 2008). Boto (2010) has reviewed the role of such horizontal gene transfers in the evolution of prokaryotes and well as unicellular eukaryotes. In prokaryotes (*E. coli*) the transfer of genes via a plasmid vector was recently demonstrated in vivo in the laboratory (Babić, Lindner, & Vulić, et al., 2008). There is now compelling evidence to support the once contentious view that HGT provides an important source of new genes and functions to recipient organisms and a driving force for evolution. The operation of HGT has significantly foiled attempts to reconstruct ancient phylogenetic relationships in search of a Last Universal Common Ancestor (LUCA) (Doolittle, 1999; Jain, Rivera, Moore, & Lake, 2003). Fiure. 2 shows the tangled connections emerging from phylogenetic analysis near the base of the "tree of life," suggesting that there is no true tree of life defined by species, only a tangled web of gene connections (Doolittle, 1999). This clearly shows that there was no Last Universal Common Ancestor (LUCA) on the Earth, but only a universal ensemble of genes.

From data currently available we might infer that sudden shifts in evolution, the emergence of new traits and even the arrival of new species, occurs through HGT rather than by the slow neo-Darwinian process of mutations and natural selection (Keeling & Palmer, 2008). Although the occurrence of neo-Darwinian gradualist evolution is not denied, it would probably be dwarfed by HGT in the long term. The phenomenon described as "punctuated equilibrium," where long periods of evolutionary stagnation are punctuated by sharp episodes of innovation and progress (Eldredge & Gould, 1972; Gould & Eldredge, 1977; Hoyle & Wickramasinghe, 1981), is consistent with cosmically mediated HGT. The long periods of slow evolution are due to Earth-bound neo-Darwinian processes, sped up in bursts by environmentally driven rapid Lamarckian adaptations (Chapter 3, Steele et al., 2019).

The successful transfer of genetic information from one organism to another in a manner that permits transmission in a host's germ line requires a vector. The vector could take the form of a plasmid, virus or a bacterium, and the host and vector is required to enter into some form of symbiosis. Eukaryotes possessing mitochondria or chloroplasts provide living evidence of horizontal gene transfers that happened in the past, with mitochondria and chloroplasts being prokaryotic endosymbionts (Keeling & Palmer, 2008). In the next section we shall cite evidence from human genome studies that similar symbiotic accommodation of viral or bacterial genes have occurred repeatedly in the primate ancestral line that led eventually to *Homo sapiens*.

We conclude this section by reiterating that the evidence that now exists for purely Earth-based gene transfers can be extended to include transfers over a galactic scale, if the biosphere within which life originates and evolves is regarded as having cosmic dimensions. Transfers of alien genes would take place whenever the solar system (and Earth) encounters genetic material (viruses and bacteria) from comets or planetary systems from which such material had been expelled. The web of life displayed in the phylogenetic map of Fig. 1 is then the product of local as well as

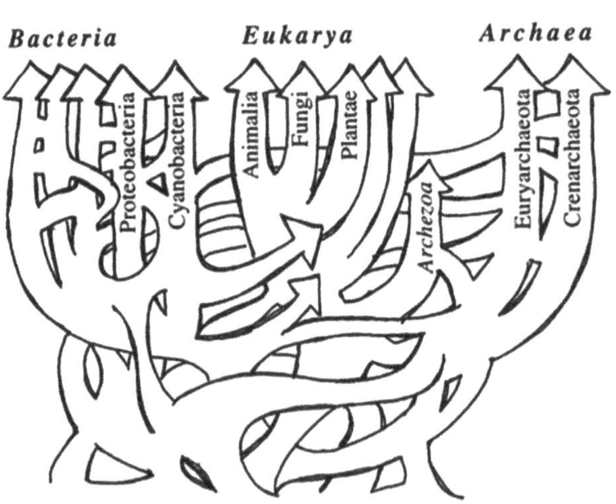

Fig. 1 Gene sequence studies reveal a tangled web of life, no tree of life, nor a convergence to a Last Universal Common Ancestor on Earth. *Adapted from Doolittle, W. F. (1999). Phylogenetic classification and the universal tree. Science, 284, 2124–2128.*

galactic "horizontal gene transfer" and we add the Lamarckian dimension to the adaptation process as a now necessary factor in the efficient cosmic spread of life (Chapter 3, Steele et al., 2019). Dynamical routes for such transfers on a galactic scale have been identified and it is to be expected that they occurred on a stochastic basis throughout the length of time for which life has existed in the Galaxy (Wallis & Wickramasinghe, 2004; Napier, 2004; Wickramasinghe, Wickramasinghe & Napier, 2010).

6. Viral sequences in genomes

Sequencing the human genome has been one of the most important scientific achievements in the new millennium. It has led to a wide range of discoveries in fundamental biology as well as medical science that are transforming our ideas about viruses, disease and evolution (Venter, Adams, Myers, et al., 2001). One surprise to follow was that the number of sequences coding for proteins (genes) is as small as 20,000–25,000 rather than over 100,000 as had earlier been suspected. The other surprise was that 8% of our genomes consist of endogenous retroviruses—RNA viruses that reverse transcribe their RNA into DNA. Their significance in causing disease as well as in contributing to evolution is only just beginning to be understood, and many astounding correspondences with statements quoted in an earlier section cannot be overlooked.

New evidence from genome studies points to frequent episodes of retroviral infections (of which HIV is an example) in almost all mammalian species (Dawkins et al., 1999; Ryan, 2004). De Groot, Otting, Doxiadis, et al. (2002) have identified an entire repertoire of genes (MHC class 1 genes) in chimpanzees that confer immunity against chimpanzee-derived simian immune deficiency virus. The inference is that modern chimp populations represent descendants of the survivors of a HIV-like pandemic that nearly culled the entire ancestral chimp populations in the distant past.

Following the integration of an offending retroviral gene sequence into a host's DNA, random mutations as well as development of host immunity within a few generations leads to the cessation of infectivity. Survivors of major pandemics thus carry DNA of retroviral origin bearing testimony to a history of prior infections (Dawkins et al., 1999; Ryan, 2004). Such endogenization of retroviruses often leads multiple inserts of the same viral sequence across the entire length of the host genome. The viral genes thereafter add to the potential for evolution in the long term through the action of point mutations and natural selection.

Studies of Horie, Honda, Suzuki, et al. (2010) have shown that the process of endogenization of viruses is not confined to retroviruses. A non-retroviral RNA appears to have been incorporated in the germ line of several mammalian species, including rodents around 40 million years ago (Horie et al., 2010). With nearly half the human genome comprised of viral sequences, it would appear reasonable to suppose that evolution of hominids leading to *Homo sapiens* involved a succession of viral or bacterial infections followed by endogenization and incorporation of novel gene sequences in the evolving germ line.

In a recent paper Wang et al. (2012) have shown that two immunomodulatory genes (SIGLEC) are inactive in humans, but not in related primates. It is conjectured that these genes when they were fully active could have been targets for a lethal bacterial infection that nearly culled the human population in the past, perhaps 100,000 years ago. Modern humans, on this conjecture, are descendants from the handful of survivors from a lethal bacterial pandemic.

The stochastic nature of evolution—long periods of quiescence punctuated by sudden bursts—as is evident in the fossil record is consistent with the introduction of transformative genes on occasions when the solar system came within reach of other life-bearing planetary systems and sources of novel genes. Such a picture is consistent with the predictions of the cosmic evolution model.

7. Further predictions of interstellar panspermia

Microorganisms on the Earth display an incredible range of functional diversity. Particular microbial species can colonize particular niches, tolerating extremes of temperature, acidity, salinity, desiccation as well as high levels of ionizing radiation. Microbial habitats include subterranean ocean caves, hydrothermal vents, polar ice and tropospheric clouds. Throughout this vast range of conditions an amazing biochemical unity prevails pointing unerringly to a single origin. According to interstellar panspermia that event was external to our solar system.

If life is discovered on planetary habitats outside the Earth such life will be expected to possess the same unity of biochemistry that is found on the Earth—the same genetic code, the same DNA, and identical biochemistries. Only the habitat will select what is best suited for survival, similar to the microorganisms inhabiting the diverse range of habitats on Earth. If independent origination events occur such correspondences will be impossible. A radically different biochemistry discovered elsewhere could be considered as a falsification of our basis premise.

Finally, if we adopt the currently accepted concordance Big Bang model of the universe, our observable horizon is at a radius of 46 billion light years. The true extent of the universe of course depends on the precise spatial curvature of the universe and constraints from MWB radiation observations point to the universe being flat to within 0.4%. The universe may therefore be at least a factor of a few hundred larger, and may even be infinite in extent (B. Klyce, *Foreword*) if it turns out that the universe is precisely flat. If life originates as a highly improbable event as is implied by present arguments, it is likely that these events will be localized within disjoint "life-domains" in the universe. The cosmological domain we happen to find ourselves in defines the nature of our idiosyncratic cosmic biology, including its basic biochemistry as well as its full range of genetic diversity—a range that is only incompletely expressed on the Earth.

8. Concluding remarks

Many biologists still tend to dismiss the theory of cometary panspermia as a "controversial idea" lacking in supportive evidence. As we have seen from the chapters here, this is far from being the case—the evidence is wide-ranging, multifactorial and specific. One might as well ask the question: what evidence is there for the primordial soup theory and the idea that life originated on the Earth? There is certainly no direct evidence in its favor, and this theory depends for its security on an obsolete pre-Copernican philosophy. In contrast, evidence from astronomy, geology and biology continue to provide strong support for the theory of cometary panspermia (Wickramasinghe, 2010); and it should be stressed that much of this evidence was obtained long after our theory was formulated, as we saw in the present chapter.

In accordance with the scientific methodology pioneered by philosophers in the 17th century we can use the feedback loop of Fig. 2 to generate cycles of prediction–verification–re-affirmation to put our theory to ever more stringent tests. Needless to say there has been a veritable list of successes and confirmations over the past three decades. The loop of Fig. 3 has been enormously strengthened and amplified in recent years. The spectroscopic

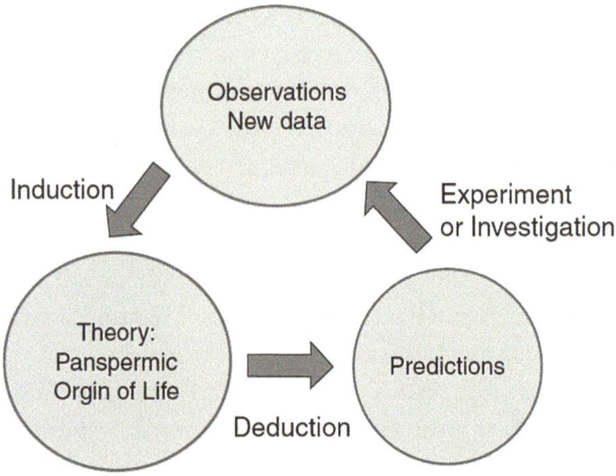

Fig. 2 The deductive/inductive method of science applied to panspermia theory.

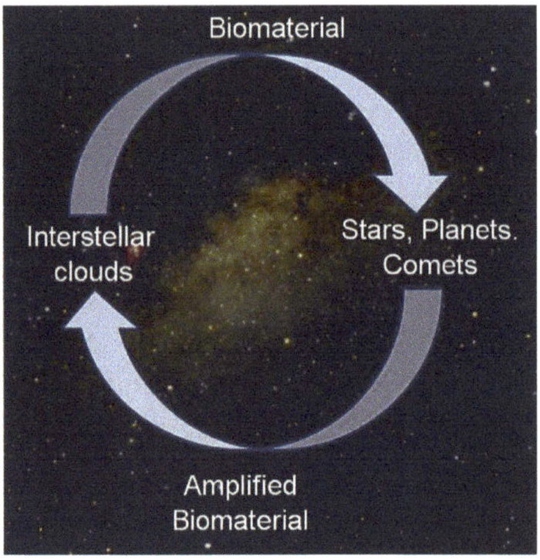

Fig. 3 A feedback loop of panspermia. Comets amplify and distribute bacteria, eukaryotic cells, viruses, e.g., see discussion of this figure also in Steele et al. (2019).

identification of interstellar dust and molecules in space as representing the detritus of biology, which was our starting point in the 1970s, has come into much sharper focus (Wickramasinghe, 2010). The biochemical relevance of interstellar material is now widely conceded, although a fashion remains to assert without proof that we are witnessing the operation of "prebiotic chemical evolution" on a cosmic scale. If biological evolution and replication are regarded as the only reliable facts, life always generates new life, and this must surely be so even on a cosmic scale. Prebiology, whether galactic or planetary, remains an unproven hypothesis that fails any test based upon a loop like Fig. 2. Prebiotic evolution implying a ready transition to life everywhere in the galaxy is in the author's view a mistaken remit of modern astrobiology and is likely to lead us astray. This is becoming ever more pertinent with the discovery of habitable exoplanets such as Kepler 22b, where the presence of biochemicals, when found, should properly be interpreted as evidence of extraterrestrial biology rather than prebiology.

In this chapter we pointed to the correspondences between many predictions of a model of cosmic biology, proposed over four decades ago, and modern developments of gene sequencing, including the human genome project. Normal conventions in science demand that such well-documented predictions, when they come to be verified subsequently, will be acknowledged as vindication of the theory. But in the present instance the connection is not even mentioned in passing and possible reasons for this were discussed earlier in this chapter. As long as the long overdue paradigm shift away from Earth-centered-origin-of-life theories remains culturally unacceptable, the full significance of horizontal gene transfer and the prevalence of viral genes in our genomes, and a universal DNA/RNA/protein paradigm with an invariant genetic code, will remain a puzzle.

References

Babić, A., Lindner, A. B., Vulić, M., et al. (2008). Direct visualization of horizontal gene transfer. *Science, 319*, 1533–1536.

Boto, L. (2010). Horizontal gene transfer in evolution: Facts and challenges. *Proceedings of the Royal Society B, 2010*(277), 819–827.

Burke, E. (1849). *The works of Edmund Burke, with a memoir 2*. Harper & Brothers, p.248.

Crick, F. H. C., & Orgel, L. E. (1973). Directed panspermia. *Icarus*, *19*, 341–346.

Dawkins, R. L., Leelayuwat, C., Gaudieri, S., Tay, G., Hui, J., Cattley, S., et al. (1999). Genomics of the major histocompatibility complex: Haplotypes, duplication, retroviruses and disease. *Immunological Reviews*, *167*, 275–304.

De Groot, N. G., Otting, N., Doxiadis, G. G. M., et al. (2002). Evidence for an ancient selective sweep in the MHC class I gene repertoire of chimpanzees. *Proceedings of the National Academy of Sciences of the United States of America*, *99*, 11748–11753.

Doolittle, W. F. (1999). Phylogenetic classification and the universal tree. *Science*, *284*, 2124–2128.

Eldredge, N., & Gould, S. J. (1972). Punctuated equilibria: An alternative to phyletic gradualism. In T. J. M. Schopf (Ed.), *Models in paleobiology* (pp. 82–115). San Francisco: Freeman Cooper.

Gould, S. J., & Eldredge, N. (1977). Punctuated equilibria: The tempo and mode of evolution reconsidered. *Paleobiology*, *3*, 115–151.

Horie, M., Honda, T., Suzuki, Y., et al. (2010). Endogenous non-retroviral RNA virus elements in mammalian genomes. *Nature*, *463*, 84–87.

Hoyle, F., & Wickramasinghe, N. C. (1979). *Diseases from space*. London: J.M. Dent & Sons.

Hoyle, F., & Wickramasinghe, N. C. (1981). *Evolution from space*. London: J.M. Dent & Sons.
Hoyle, F., & Wickramasinghe, N. C. (1990). Influenza—Evidence against contagion. *Journal of the Royal Society of Medicine, 83*, 258–261.
Jain, R., Rivera, M. C., Moore, J. E., & Lake, J. A. (2003). Horizontal gene transfer accelerates genome innovation and evolution. *Molecular Biology and Evolution, 20*(10), 1598–1602.
Keeling, P. J., & Palmer, J. D. (2008). Horizontal gene transfer in eukaryotic evolution. *Nature Reviews Genetics, 9*, 605–618.
Liu, Y. (2018). Darwin's pangenesis and graft hybridization. *Advances in Genetics, 102*, 27–66.
Napier, W. M. (2004). A mechanism for interstellar panspermia. *Monthly Notices of the Royal Astronomical Society, 348*, 46–51.
Ryan, F. P. (2004). Human endogenous retroviruses in health and disease: A symbiotic perspective. *Journal of the Royal Society of Medicine, 97*, 560–565.
Steele, E. J., Gorczynski, R. M., Lindley, R. A., Liu, Y., Temple, R., Tokoro, G., et al. (2019). Lamarck and Panspermia-on the efficient spread of living systems throughout the cosmos. *Progress in Biophysics and Molecular Biology, 149*, 10–32.
Venter, J. C. J., Adams, M. D., Myers, E. W., et al. (2001). The sequence of the human genome. *Science, 291*, 1304–1351.
Wallis, M. K., & Wickramasinghe, N. C. (2004). Interstellar transfer of planetary microbiota. *Monthly Notices of the Royal Astronomical Society, 348*, 52–57.
Wang, X., Mitra, N., Secundino, I., Banda, K., Cruz, P., Padler-Karavani, V., et al. (2012). Specific inactivation of two immunomodulatory SIGLEC genes during human evolution. *Proceedings of the National Academy of Sciences of the United States of America, 109*, 9935–9940.
Wesson, P. (2010). Panspermia, past and present: Astrophysical and biophysical conditions for the dissemination of life in space. *Space Science Reviews, 156*(1–4), 239–252.
Wickramasinghe, N. C. (2010). The astrobiological case for our cosmic ancestry. *International Journal of Astrobiology, 9*, 119–129.
Wickramasinghe, N. C. (2011). Viva panspermia!. *Observatory Magazine,* April 2011.
Wickramasinghe, J. T., Wickramasinghe, N. C., & Napier, W. M. (2010). *Comets and the origin of life*. WSPC.
Wickramasinghe, J. T., Wickramasinghe, N. C., & Wallis, M. K. (2010). *Comets and the origin of life*. Singapore: World Scientific.

Further reading

Hoyle, F., & Wickramasinghe, N. C. (2000). *Astronomical origins of life*. Dordrecht: Kluwer Academic Publ.
Kwok, S. (2009). Organic matter in space: From stardust to the solar system. *Astrophysics and Space Science, 319*, 5–21.
Trevors, J. T., Pollack, G. H., Saier, M. H., Jr., & Masson, L. (2012). Transformative research: Definitions, approaches and consequences. *Theory in Biosciences, 131*(2), 117–123. https://doi.org/10.1007/s12064-012-0154-3.
Wainwright, M., Wickramasinghe, N. C., Narlikar, J. V., & Rajaratnam, P. (2003). Microorganisms cultured from stratospheric air samples obtained at 41km. *FEMS Microbiology Letters, 218*, 161.

CHAPTER FIVE

The mutagenic source and power of our own evolution

Robyn A. Lindley*
GMDx Group Ltd, Melbourne, VIC, Australia
Department of Clinical Pathology, Faculty of Medicine, Dentistry & Health Sciences, University of Melbourne, Melbourne, VIC, Australia
*Corresponding author: e-mail address: robyn.lindley@unimelb.edu.au

Contents

Abstract

This chapter addresses the molecular mechanism and source of the numerous mutagenic changes that genomes, particularly mammalian genomes, can undergo during normal development and in diseased states. The central role of pathogen and other disease-inducing innate immunity via the action of the cytosine (AID/APOBEC) and adenosine (ADAR) deaminases is reviewed in some depth. The general and universal nature of deaminase-mediated mutagenesis is an important key to understanding cosmic genetic evolutionary processes.

1. Introduction

Humanity is going to require a new way of thinking if it is to survive.

Albert Einstein

As we enter a new evidence-based era driven by low cost whole genome sequencing and big data analytics, our understanding about genomics and the generation of genetic diversity is forcing us to ask new questions about

Advances in Genetics, Volume 106
ISSN 0065-2660
https://doi.org/10.1016/bs.adgen.2020.05.001

the mutagenic source and power of our own cosmic evolution: Once carbon-based life as we know it was seeded, what role do endogenous mutagenic factors play in adaptation, survival and in propelling genetic diversity across species?

We now know that the main endogenous source of genetic diversity is due to the mutagenic activity of enzymes called deaminases (Chapter "The Efficient Lamarckian Spread of Life in the Cosmos" by Steele et al., table 1). But why is life on Earth pre-packaged with deaminases and the molecular machinery required for survival, adaptation and evolution? Are deaminases the "universal mutators" responsible for our own evolution? The aim of this chapter is to introduce the role of deaminases in health and disease, and to understand them as the source and power assisting our own genetic evolution.

2. Deaminases as an endogenous source of mutations

Whatever the source of carbon-based life on earth, we know that most DNA-RNA life forms carry a cargo that includes genes encoding deaminases. From yeast to man, deaminases have now been found in the genetic material of most animal species. Our growing understanding of the mutational role of deaminases has resulted in a paradigm shift away from the idea that mutations arise randomly. While some mutations are generated directly by external physical sources such as ionizing radiation and hazardous chemicals, the deaminases are truly endogenous.

Deaminases are activated in all cells, mainly by invading pathogens, viruses, bacteria and fungi (Schoggins & Rice, 2011). Thus, they play a pivotal role in the cause of gene mutations coupled to environmental signals. Together, these two processes support an updated view of adaptive evolution via endogenous genetic mechanisms involving epigenetic changes and deaminase mutational activity across the genome. The "Darwinian" roles of cell death and organism survival will continue to ensure that *natural selection* also plays a part in choosing those cells or individuals that will survive long enough to produce progeny of their own. However, the source and power of our own evolution is the deaminases. Understanding this also draws each of us closer to understanding our own existence.

In humans, there are around 14 different deaminase proteins that change the structure of a gene by altering a single base unit in our DNA (reviewed in Mamrot, Balachandran, Steele, & Lindley, 2019). New mutations are observed in somatic tissue and are the result of deaminases targeting genes for deamination. If the resulting mutational change remains uncorrected by DNA or RNA repair mechanisms, the result is a new somatic mutation in the DNA of that cell. For example, the result may be the mutation of an "A" (adenosine) to a "G" (guanosine) in a particular gene.

Most deaminases only target a specific tissue or tissue group. These processes are intrinsically linked to transcription. That is, new mutations occur on genes when they are being actively transcribed and implicate the nascent RNA as a copying template intermediary. Thus, the information flow is: "DNA-to-pre-mRNA-to-DNA." This provides an opportunity for de novo mutations to be introduced into DNA or RNA before a modified nucleotide sequence is copied into newly transcribed DNA via a reverse transcription step. A simplified figure showing this process is provided in Fig. 1.

There are only two types of deaminase families: those targeting cytidines (Cs) that result in mutations of Cs, and those targeting adenosines (As). The cytidine deaminases target single stranded DNA (ssDNA) which usually only occur in the context of an "open transcription bubble." The cytidine deaminase family includes AID, APOBECs 1, 3A, 3B, 3C, 3D, 3F, 3G, 3H, and result in uracil bases (C-to-U) which are mutagenic in DNA and subject to base excision repair by Uracil DNA glycosylase (UNG) often resulting in Abasic sites (AP). Uracils and AP sites are thus the primary AID/APOBEC deaminase lesions in DNA requiring to be repaired by DNA repair pathways.

Adenosine deaminases (ADARs 1 and 2) target adenosines in double stranded RNA (dsRNA) that occur when RNA snaps back on itself to form complex double strands and loop structures that are often referred to as "hairpins." Recently it has also been discovered that the adenosine deaminases also target hybrid DNA-RNA structures that are created as the newly formed mRNA template is spun off the bottom strand (transcribed strand) of the DNA in the transcription bubble. Hybrid DNA-RNA structures may also be targeted by adenosine deaminases during the reverse transcription (RT) step in which the newly formed DNA template is formed from an intermediary RNA template (Steele & Lindley, 2017; Zheng, Lorenzo, & Beal, 2017).

Each deaminase targets a specific sequence motif that defines the inferred deaminase binding domain (InfDBD). That is, we can identify the likely deaminase source of mutations by using DNA sequence analyses to identify the preferred motif (nucleotide sequence) that is interpreted as the InfDBD biochemically complementary target site, the type of resulting mutation and

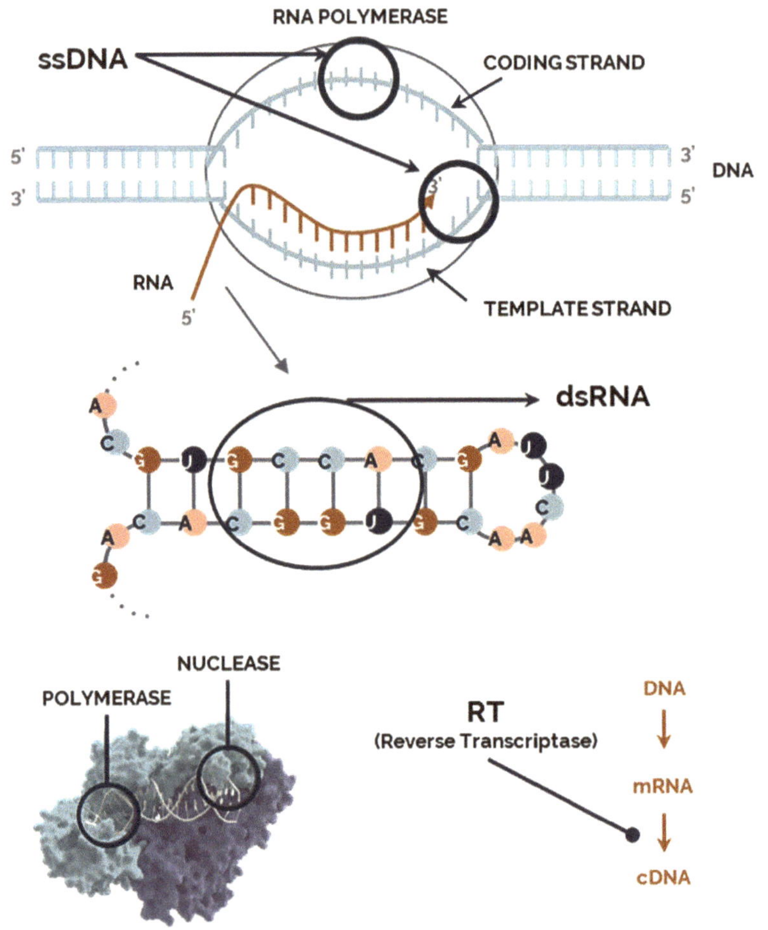

Fig. 1 Schematic diagram showing the transcription process that is initiated in the open transcription bubble (constructed from concepts in Lindley, 2018). The newly templated RNA intermediary is produced from the transcribed strand (bottom strand) template and forms a snapback hairpin structure that results in dsRNA. During reverse transcription (RT) an intermediary RNA template is used to produce the newly formed cDNA. Uncorrected new mutations may be generated in the transcription bubble, the dsRNA or during the RT step (see also Steele & Lindley, 2017, reviewed in Franklin, Steele, & Lindley, 2020).

the preferred target site in protein-coding regions that in turn define any resulting changes in the new amino acid sequence. For example, one typical AID deaminase signature resulting from the uncorrected deaminase activity of AID is described as: a C-to-T mutation, off a WRC motif (where R=A/G, W=A/T), and preferentially occurring at a mutated codon target site MC1 (Mutated Codon, read in the 5-prime to 3-prime direction, Lindley, 2013). Deaminase mutational activity predominantly produces transition mutations defined as A-to-G, T-to-C, C-to-T or G-to-A. To increase the number of possible mutation target sites during transcription, each deaminase may also have more than one InfDBD, and some deaminases can self-edit their InfDBDs. Thus, together deaminases provide a rich source of potential new mutations of all types, and in all genes that are expressed.

3. Deaminases in immunity

The deaminases provide the main endogenous source of potential new genetic variants required to generate diversity in immune system genes, and often other genes during transcription, and as evidenced by the genome-wide somatic mutation signatures of cancer genomes (Lindley, 2013; Lindley, Humbert, Larmer, Akmeemana, & Pendlebury, 2016). In the immune system, a newly mutated variable (V-) gene is "clonally selected" in a true Darwinian manner if it produces the "best fitting" antibodies to protect against the disease. This step is also Lamarckian in the sense that the invading pathogen from the "environment" provides the trigger for initiation of the cascading immune response. Some of these new strongly selected somatic antibody V mutants are predicted to be transferred back to the germline for the benefit of future generations (Lindley, 2011a, 2011b, 2018; Steele, 1979; Steele & Lindley, 2018; Steele, Lindley, & Blanden, 1998). It has been known for almost two decades that the antibody V gene clusters are dependent upon deaminase mutagenic activity to regularly update the antibody repertoire of an individual as a part of the body's immune response to foreign microorganisms (Steele & Lindley, 2018). These updates may also be inherited. In the case of smallpox or polio for example, previously exposed generations might then confer, *in part*, some genetically acquired resistance to future generations (e.g., see Steele & Lloyd, 2015).

4. Deaminase target site access is directed by epigenetic markers

Current research on the origins of all types of mutations is now focused on understanding the role of deaminases in normal somatic cells during disease. The new molecular evidence suggests a far more general role for deaminase-based mutator mechanisms than was originally discovered for V gene updates in response to invading pathogens. This evidence is supporting the idea that environment-driven genetic and epigenetic changes, which when combined, target new sites for non-random mutation in many other genes across our genome. It is this genetic-epigenetic coupling that now underpins new evolutionary thought (this concept is reviewed in depth in Steele et al., 2019, see fig. 2).

It has been known for over two decades that a number of processes write additional "regulatory" information onto the surface of the genes without altering the genome sequence, a process termed "epigenetic" or "soft" re-wiring of the genome (Jablonka & Lamb, 1995). These epigenetic events make "surface footprints" on the gene sequences (composed of A,T,C,G nucleotides) making up a part or all of the genetic regions targeted. Many epigenetic changes triggered by the environment have been shown to be stably inherited by offspring for several generations (Jablonka & Lamb, 1995; Lindley, 2010, 2011a, 2011b, 2018).

What has been discovered since then is that environmentally triggered epigenetic markers also target genomic regions for deaminase action (Lindley, 2018; Scourzic, Mouly, & Bernard, 2015), and to protect those genes and regions of genes that are to be conserved. That is, environmentally driven epigenetic changes are used to mark which genes or parts of genes are potential deaminase targets sites where it might be beneficial for an organism to have some DNA sequence updates.

Thus, environmentally triggered epigenetic changes *directly mark* potential gene sites for possible new gene mutations or conservation. Referring to Fig. 1, a schematic representation of the linkages implied by these coupled events is shown. Scourzic et al. (2015) have reviewed many of the potential molecular steps we think are important in these environmentally triggered epigenetic-genetic mutation paths.

Hence it is important to realize that, deaminases alone are not directional drivers of mutations. They are merely the tools with which the biome relies upon to introduce and regulate deamination events that may lead to new genomic variants. In Fig. 2, a schematic diagram shows the main links between environmental triggers, gene regulation and expression paths, and the use of epigenetic markers to identify gene regions that are potentially accessible to deaminases.

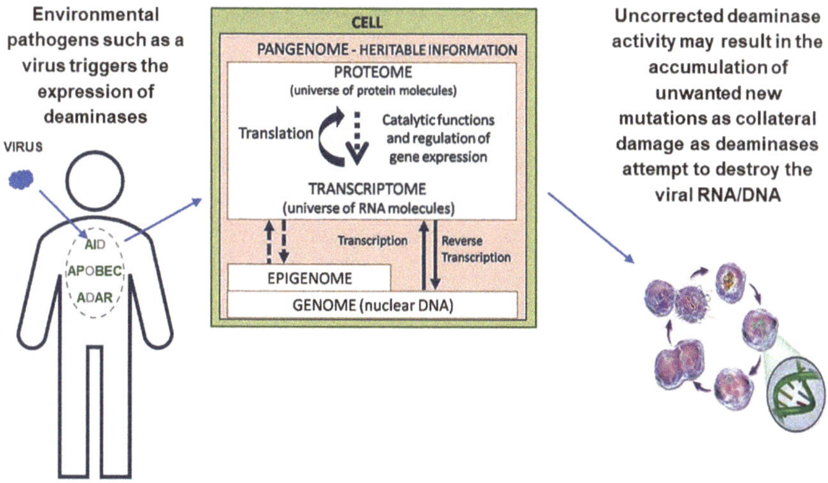

Fig. 2 Schematic diagram showing the main links between the environment and the proteome. The proteome links epigenetic changes to targeted deaminase mutagenic activity. The accumulation of unwanted new mutations in daughter cells of the target tissue(s) may result in abnormal cell growth, and a diagnosis of cancer. Constructed in part from elements in Lindley (2018).

5. Linking deaminase activity to the formation and inheritance of instincts

A key question is: How do natural instincts arise in evolution? Ever since 1887 with Samuel Butler's *Luck or Cunning?* simple logic for many has implied that strong survival instincts based on fear responses *must have arisen* in our ancestors—not by chance, but in a Lamarckian manner which were then passed on to their progeny. A recent report by Dias and Ressler (2014) bears directly on how instinctual responses arise. Parental mice were subjected to Pavlovian odor fear conditioning before conception. It was then observed that subsequent generations had an increased specific behavioral sensitivity to the specific chemical odor used to condition the parents. That is, unrelated chemical odors did not trigger a fear response. They also did other breeding experiments that established that these specific acquired transgenerational effects are indeed inherited via parental gametes. Thus, both direct genomic and indirect epigenetic odorant receptor gene targeting *appear to act together* to establish what we now recognize as specific instinctual responses involving odorant receptor genes and behavior. We do not know all the molecular steps involved, but current data suggests that environmental triggers direct specific families of proteins and associated molecules to introduce the epigenetic changes needed to open up specific genes for enhanced expression, and to identify the part of a gene that is potentially available for mutation by deaminases. These epigenetic changes then permit susceptibility to deaminase action that may lead to permanent genetic change (Lindley, 2018; Scourzic et al., 2015, and see fig. 2 in Steele et al., 2019).

It is this new understanding of epigenetic-genetic coupling that is required to direct the location of new genetic variants. It also helps us to better explain some other previously reported Lamarckian results. The now classic Lamarckian experiments of 1918–24 by Guyer and Smith (data republished with modern interpretation in Steele, 2016) on maternally induced eye defects in rabbits can be viewed as a complex combination of the antibody V gene soma-to-germline feedback process, and transgenerational epigenetic modifications involving both genomic access and targeted gene expression (Scourzic et al., 2015). This new reality also reminds us of the work of Frederic Wood Jones. He emphasized many years ago (Wood Jones, 1943) that coupled associated experiences together with acquired habits are powerful drivers of organic change.

As these and some other Lamarckian-like occurrences can now be explained using molecular genetics, many mainstream biologists have in recent years openly supported the idea of Lamarckian modes of inheritance as a key mechanism underpinning our current view of evolution (Lindley, 2018; Noble, 2013, 2020; Noble et al., 2016; Skinner, 2015).

6. Deaminases in disease

It is now recognized that the development of many cancers and cancer progression itself is the result of dysregulated deaminase activity (Lindley et al., 2016; Petit, Vartanian, & Wain-Hobson, 2009). As uncorrected mutations are introduced to DNA, the result is aberrant protein production that may result in a diagnosis of cancer as shown in Fig. 1.

As well as causing genetic damage to our body cells that could result in a diagnosis of cancer, there is also very recent evidence that the deaminase mutators have helped to shape the genetic landscape of the human germline over evolutionary time. That is, alter the genomes encased in the cells of the eggs and sperm of the reproductive organs. It has been shown that the genetic variations in the human germline known as "single nucleotide polymorphisms" (SNPs, pronounced as "SNIPS") occur at specific genetic signatures associated with deaminase mutational activity (Lindley & Hall, 2018). This suggests that the original source of many SNPs may be the result of deaminase-linked mutation processes, and that these have been inherited from ancestors, and continue to persist in human sub-populations. The environmental trigger for some of the SNPs arising in our ancestors genes could be the result of deaminase anti-viral activity. Some deaminases are widely known as "virus smashers." Unfortunately, in the process of their mutational action intended to kill and destroy a virus, our genes may also take an unintended mutational "hit" as collateral damage (Petit et al., 2009). While lethal SNPs may arise, only a small number of SNPS cause lethal genetic diseases. Most are more benign highly adapted modifications of protein structure and function. The finding that many human SNPs occur at known deaminase genomic signatures is difficult to explain without the invocation of a soma-to-germline feedback loop as part of their primary cause in a previous generation (Lindley & Hall, 2018).

As far back as 2011, it was speculated that…"It may well be that the SHM mechanism (viz. somatic hypermutation in antibody V genes) is an important mechanism that is co-opted in some way to update the DNA of a large number of genes in healthy tissues and germline cells" (Lindley, 2011a). It now appears that this speculation has since been confirmed by analyzing mutation patterns in thousands of non-immune system genes across both cancer genomes (Lindley, 2013; Lindley et al., 2016), and the germline genomes of patients suffering a wide range of inherited genetic diseases that we now associate with "SNP" tests—some serious, the majority somewhat benign (Lindley & Hall, 2018).

7. The role of deaminases in normal healthy cells

What function do the deaminases perform in normal healthy cells? They have been recognized as potentially dangerous mutators for some time. It would have to be a regulated process of targeted somatic "mutation" that is beneficial to the cell and organism, otherwise cancer might easily develop. We know that the different types of deaminases are expressed to variable levels across the different tissues of the body (Koning et al., 2009; Refsland et al., 2010). During pathogenic virus infections such as the flu or HBV (hepatitis B virus) infection, the deaminase-mediated mutations literally riddle the viral genes with lethal mutations (Petit et al., 2009). In a normal healthy person, many of the mutations occurring in the cell's genome are repaired to limit the number of unwanted new mutations being introduced to the genome as "collateral damage" while fighting an infection such as is shown in Fig. 2. Clearly, a healthy person needs to be able to mount an effective deaminase response to viral infection, and then ensure that the mutations are repaired for the next generation of cells—or offspring. One would therefore expect and hope that these processes imply that the action of deaminases themselves is safely regulated in the various tissues of a healthy individual.

Thus, there has been a recent cut-through in the tacit acceptance of the likely deaminase origins of gene mutations that imply a distinct Lamarckian-like process. However, many of the molecular mechanisms underpinning the intergenerational transfer of new genetic cannot at present be fully explained (Steele et al., 2019).

8. The role of big data

The influx of massive quantities of whole genome DNA sequence information via Next Generation Sequencing technology (NGS) has been decisive. It is the overwhelming driver in this changed conceptual understanding. We now confidently talk of the "genomic landscape in cancer" (Vogelstein et al., 2013), and this is now applied on a significantly wide scale in the clinic. Hundreds of thousands, if not millions of whole genomes or their significant protein-coding sub-portions (the "exome" component encompassing 2% of the genome) have now been sequenced.

The NGS technological revolution has therefore utterly changed all aspects of biomedical research, particularly clinical personalized medicine. Numerous "big data" algorithms and bioinformatics pipelines come to the fore. Indeed deaminase-specific algorithms now allow for the identification of *which* deaminase mutator is *likely to have caused* the identified somatic mutations during cancer development, or to identify the likely origin of an inherited germline mutation. The conservative estimate is that 30–40% of all somatic and germline genetic variations can be accounted for by just four or so members of the 14 known human deaminases.

9. Consciously directing the evolution of HomoSapv2.0

At this historical juncture, learning to control the unintended mutational consequences of a deaminase response to an invading pathogen forces us to renew our thinking about how to manage health and disease, and thus direct our evolutionary future for "wellness." It is therefore important to further develop our knowledge of the complex interplay between the environment and evolution, and to understand the likely deaminase source of new somatic mutations as cancer progresses so that we may prevent cancer in the future. We also need to develop our understanding of how de novo deaminase associated mutations are inherited in the germline. Together, these new learnings will continue to reveal just how important the role of deaminases is in understanding our own health, the health of our ancestors, and our relationship with the environment in which we live.

As we confront the growing amount of molecular and genetic data showing that we are designed to adapt to our environment, sooner or later we will also need to confront the personal and global responsibilities of consciously directing our own evolutionary destiny. Through the conscious choices we make, we can alter the pace and direction of our evolution. These processes may also be augmented by new genetic engineering options. This realization has the potential to propel our mass consciousness to another level altogether.

New genetic analyses are showing that our DNA can be altered by a wide range of environmental challenges such as emotional trauma, lifestyle choices like smoking and depression. We have started to become aware that our own DNA mirrored at least some of the changes occurring in our life. We are discovering that our actions and our genes were more closely linked than we ever thought possible in the past. Like pieces in a giant jigsaw puzzle, genetic and epigenetic data are gradually enabling us to see a new image of ourselves as a highly mutable living organism with the power to direct our future evolution. But what will HomoSapv2.0 look like?

The precepts of conscious evolution therefore lie in our understanding of how the environment and our lifestyle choices impact our inheritable material. What is arising is an opportunity to drive further evolutionary augmentation and greater diversity of us as a species. Never before have we had such a choice. Yet, to define a new vision so that we can complete our own image in the giant jigsaw puzzle of life is a new journey that we are only just beginning. Philosophers love this sort of dilemma. Living through a paradigm shift of this magnitude and spanning so many disciplines could potentially have a huge impact on how we perceive ourselves, our society and our moral codes. We have far more conceptual realities and possibilities to fit into our philosophical prisms.

While further research will develop our understanding of the underlying molecular processes, it will not help us to define HomoSapv2.0. Right now, as we contemplate what our evolutionary future might be, we are all actively writing the code for HomoSapv2.0. Simply by being aware of the processes involved, the genetic engineering technologies available and lifestyle possibilities that this might lead to genetic changes, we are already laying down an early blueprint for HomoSapv2.0.

References

Butler, S. (1887). *Luck or cunning?* Reissued May 25, 2017. Available Ashed Phoenix – Million Book Edition. ISBN: 9781434666451. Amazon.com: Books.

Dias, B. G., & Ressler, K. J. (2014). Parental olfactory experience influences behavior and neural structure in subsequent generations. *Nature Neuroscience*, *17*, 89–96.

Franklin, A., Steele, E. J., & Lindley, R. A. (2020). A proposed reverse transcription mechanism for (CAG)n and similar expandable repeats that cause neurological and other diseases. *Heliyon*, *6*, e03258. https://doi.org/10.1016/j.heliyon.2020.e03258.

Jablonka, E., & Lamb, M. J. (1995). Epigenetic inheritance and evolution: The Lamarckian dimension. In *1995*, Oxford: Oxford University Press.

Koning, F. A., Newman, E. N. C., Kim, E.-Y., Kunstman, K. J., Wolinsky, S. M., & Malim, M. H. (2009). Defining APOBEC3 expression patterns in human tissues and hematopoietic cell subsets. *Journal of Virology*, *83*, 9474–9485. https://doi.org/10.1128/JVI.01089-09.

Lindley, R. A. (2010). *The soma: How our genes really work and how that changes everything!* CYO Foundation. ISBN: 1451525648, POD book Amazon.com. 2010.

Lindley, R. A. (2011a). How evolution occurs: Was Lamarck also right? *EdgeScience*, 6–9. #8 July-September 2011. https://www.scientificexploration.org/edgescience.

Lindley, R. A. (2011b). Born to evolve; how mutational and epigenetic changes enable adaptive evolution. *GIT Laboratory Journal*, *3–4*(2011), 2–4.

Lindley, R. A. (2013). The importance of codon context for understanding the Ig-like somatic hypermutation strand-biased patterns in TP53 mutations in breast cancer. *Genetics*, *206*, 222–226. https://doi.org/10.1016/j.cancergen.2013.05.016.

Lindley, R. A. (2018). A new treaty between disease and evolution—Are deaminases the "universal mutators" responsible for our own evolution? *EdgeScience*, *36*, 16–20. http://www.scientificexploration.org/edgescience/edgescience-issue-36.

Lindley, R. A., & Hall, N. A. (2018). APOBEC and ADAR deaminases may cause many single nucleotide polymorphisms curated in the OMIM database. *Mutation Research, Fundamental and Molecular Mechanisms of Mutagenesis*, *810*, 33–38. https://doi.org/10.1016/j.mrfmmm.2018.03.008.

Lindley, R. A., Humbert, P., Larmer, C., Akmeemana, E. H., & Pendlebury, C. R. R. (2016). Association between targeted somatic mutation (TSM) signatures and HGS-OvCa progression. *Cancer Medicine*, *5*, 2629–2640. https://doi.org/10.1002/cam4.825.

Mamrot, J., Balachandran, S., Steele, E. J., & Lindley, R. A. (2019). Molecular model linking Th2 polarized M2 tumour-associated macrophages with deaminase-mediated cancer progression mutation signatures. *Scandinavian Journal of Immunology*, *2019*, e12760. https://doi.org/10.1111/sji.12760.

Noble, D. (2013). Physiology is rocking the foundations of evolutionary biology. *Experimental Physiology*, *98*, 1235–1243.

Noble, D. (2020). *Exosomes, gemmules, pangenesis and Darwin. Exosomes—A clinical compendium* (pp. 487–501). Elsevier.

Noble, D., et al. (2016). *New trends in evolutionary biology: Biological, philosophical and social science perspectives* (pp. 6–9). London: The Royal Society. Carlton House Terrace, London, SW1Y 5AG Nov 7 -9, 2016. Editors-Organisors.

Petit, V., Vartanian, J.-P., & Wain-Hobson, S. (2009). Powerful mutators lurking in the genome. *Philosophical Transactions of the Royal Society B, 364*, 705–715. https://doi.org/10.1098/rstb.2008.0272.

Refsland, E. W., Stenglein, M. D., Shindo, K., Albin, J. S., Brown, W. L., & Harris, R. S. (2010). Quantitative profiling of the full APOBEC3 mRNA repertoire in lymphocytes and tissues: Implications for HIV-1 restriction. *Nucleic Acids Research, 38*, 4274–4284.

Schoggins, J. W., & Rice, C. M. (2011). Interferon-stimulated genes and their antiviral effector functions. *Current Opinion in Virology, 1*(6), 519–525.

Scourzic, L., Mouly, E., & Bernard, O. E. (2015). TET proteins and the control of cytosine demethylation in cancer. *Genome Medicine*, 7(2015), 9. https://doi.org/10.1186/s13073-015-0134-6.

Skinner, M. K. (2015). Environmental epigenetics and a unified theory of the molecular aspects of evolution: A neo-Lamarckian concept that facilitates neo-Darwinian evolution. *Genome Biology and Evolution*, 7, 1296–1302.

Steele, E. J. (1979). *Somatic selection and adaptive evolution: On the inheritance of acquired characters* (1st ed.) (p. 1981). Chicago: University of Chicago Press Williams Wallace, Toronto, 1979; 2nd ed.

Steele, E. J. (2016). Origin of congenital defects: Stable inheritance through the male line via maternal antibodies specific for eye lens antigens inducing autoimmune eye defects in developing rabbits in utero. In M. Levin & D. S. Adams (Eds.), *Ahead of the curve-hidden breakthroughs in the biosciences. 2016.* Bristol, UK: Michael Levin and Dany Spencer Adams IOP Publishing Ltd (chapter 3).

Steele, E. J., Gorczynski, R. M., Lindley, R. A., Liu, Y., Temple, R., Tokoro, G., et al. (2019). Lamarck and panspermia—on the efficient spread of living systems throughout the cosmos. *Progress in Biophysics and Molecular Biology, 149*, 10–32.

Steele, E. J., & Lindley, R. A. (2017). ADAR deaminase A-to-I editing of DNA and RNA moieties of RNA:DNA hybrids has implications for the mechanism of Ig somatic hypermutation. *DNA Repair, 55*, 1–6.

Steele, E. J., & Lindley, R. A. (2018). Germline V repertoires: Origin, maintenance, diversification. *Scandinavian Journal of Immunology, 87*, e12670. https://doi.org/10.1111/sji.12670.

Steele, E. J., Lindley, R. A., & Blanden, R. V. (1998). *Lamarck's signature: How retrogenes are changing Darwin's natural selection paradigm.* Sydney, Australia: Allen & Unwin.

Steele, E. J., & Lloyd, S. S. (2015). Soma-to-germline feedback is implied by the extreme polymorphism at IGHV relative to MHC. *Bioessays, 37*, 557–569.

Vogelstein, B., Papadopoulos, N., Velculescu, V. E., Zhou, S., Diaz, L. A., Jr., & Kinzler, K. W. (2013). Cancer genome landscapes. *Science, 339*, 1546–1558.

Wood Jones, F. (1943). Habit and heritage. In *1943*. London: Kegan Paul, Trench, Trubner & Co.

Zheng, Y. C., Lorenzo, C., & Beal, P. A. (2017). DNA editing in DNA/RNA hybrids by adenosine deaminases that act on RNA. *Nucleic Acids Research, 45*, 3369–3377.

CHAPTER SIX

Origin of new emergent Coronavirus and Candida fungal diseases—Terrestrial or cosmic?

Edward J. Steele[a,b,c,*], Reginald M. Gorczynski[d], Robyn A. Lindley[e,f], Gensuke Tokoro[g], Robert Temple[h], and N. Chandra Wickramasinghe[c,g,i,j,*]

[a]C.Y.O'Connor ERADE Village Foundation, Piara Waters, Perth, WA, Australia
[b]Melville Analytics Pty Ltd, Melbourne, VIC, Australia
[c]Centre for Astrobiology, University of Ruhuna, Matara, Sri Lanka
[d]University Toronto Health Network, Toronto General Hospital, University of Toronto, Toronto, ON, Canada
[e]Department of Clinical Pathology, Faculty of Medicine, Dentistry & Health Sciences, University of Melbourne, Melbourne, VIC, Australia
[f]GMDx Group Ltd, Melbourne, VIC, Australia
[g]Institute for the Study of Panspermia and Astroeconomics, Gifu, Japan
[h]The History of Chinese Science and Culture Foundation, Conway Hall, London, United Kingdom
[i]University of Buckingham, Buckingham, United Kingdom
[j]National Institute of Fundamental Studies, Kandy, Sri Lanka
*Corresponding authors: e-mail address: e.j.steele@bigpond.com; ncwick@gmail.com

Contents

Advances in Genetics, Volume 106
ISSN 0065-2660
https://doi.org/10.1016/bs.adgen.2020.04.002

Abstract

The origins and global spread of two recent, yet quite different, pandemic diseases is discussed and reviewed in depth: *Candida auris*, a eukaryotic fungal disease, and COVID-19 (SARS-CoV-2), a positive strand RNA viral respiratory disease. Both these diseases display highly distinctive patterns of sudden emergence and global spread, which are not easy to understand by conventional epidemiological analysis based on simple infection-driven human- to-human spread of an infectious disease (assumed to jump suddenly and thus genetically, from an animal reservoir). Both these enigmatic diseases make sense however under a Panspermia in-fall model and the evidence consistent with such a model is critically reviewed.

1. Introduction

In the past 40 years there have been a number of suddenly emerging epidemic viral diseases. Many were self-limiting and "went away" or "disappeared" almost as quickly as they appeared (SARS, MERS, ZIKAV). The origins in all cases were a mystery, and very controversial. Others such as the far more deadly HIV retrovirus has finally succumbed to highly effective antiretroviral therapy (HAART) making life bearable for infected HIV+ people. However it has integrated into the human germline in many cases and is likely to be a permanent "endogenized retroviral signature" in the human germline, joining the many thousands of other HERVS, human endogenous retrovirus sequences that litter the human genome as fragments or potentially active retroviruses (Wickramasinghe, 2012; Wickramasinghe & Steele, 2016).

However the great exemplar of the emergence of a new pandemic disease of considerable virulence and pathogenicity was the Spanish Flu Pandemic 1918–1919. That pandemic has been analyzed in great detail by Hoyle and Wickramasinghe (1979), and the astute and engaged reader of all that evidence is left *with only one conclusion*—the Spanish Flu disease came from Space on a massive scale, and killed tens of millions before the advent of air travel.

We do not intend here to discuss these earlier epidemics and pandemics—which have been covered in previous papers (some cited here). We focus our analysis on the actual origins of two recently emergent epidemics: a fungal disease caused by *Candida auris* and the current coronavirus "common cold-type" epidemic caused by the COVID-19 virus. These two epidemics display distinctive features and clear evidence that they may have come from a space in-fall of infectious viruses and micro-organisms in cometary dust or meteorite-derived dust particles.

2. Sudden simultaneous emergence of *Candida auris* infections in separate global regions

Candida species are well-known yeasts that can cause a variety of cutaneous and invasive infections; however, they had never been considered a serious global health threat until the recent emergence of *Candida auris*. This was first reported in the ear canal of a patient in Japan in 2009. Since then, cases have been recorded on all continents except Antarctica (Rhodes & Fisher, 2019). It can cause a variety of invasive infections with a mortality rate of up to 60%, typically infecting susceptible hosts, namely those with long hospitalizations, many illnesses and impaired immunity (Bradley, 2019). In addition, it can be resistant to multiple antifungals and has the capacity to cause outbreaks within healthcare facilities (Chow et al., 2018). Its ability to colonize and persist for a long time on human skin, tolerate some disinfectants that are commonly used in healthcare settings, and to survive on inanimate surfaces for many weeks, all contribute to its effectiveness as an outbreak agent (Jackson et al., 2019).

Even more remarkable though is its emergence. An analysis of whole genomic sequencing from 54 isolates of *C. auris* from four regions around the world revealed four major clades or genetically distinct populations. This finding supports the hypothesis of the nearly simultaneous and independent emergence of these clades in geographically separate human populations The SENTRY Antifungal Surveillance Program is a global system that has continued for 20 years (1997–2016). It collects consecutive invasive *Candida* isolates from medical centers located in four regions during each calendar year, namely: North America, Europe, Latin America, and the Asia-Pacific (Pfaller, Diekema, Turnidge, Castanheira, & Jones, 2019). Despite going back to 1997, the SENTRY data did not identify *C. auris* until 2009 (Jackson et al., 2019). In fact, the earliest *C. auris* isolates were found in South Korea in 1996 and Japan in 1997 (Forsberg et al., 2018).

Although it is a Candida species, *C. auris* is quite distinct from its Candidal relatives. The genus consists of >500 species, many of which greatly differ from each other. *C. auris* comes from the Clavispora clade of the Metschnikowiaceae family. It has not been identified from any natural environments (Jackson et al., 2019). It is relatively thermotolerant in that it can grow at temperatures as high as 42 °C. Such thermotolerance could potentially allow it to infect avian hosts (Chatterjee et al., 2015).

Thus, infections caused by the fungus *Candida* spp. have been recognized for many years. However, of interest here is the abrupt emergence of a new strain *Candida auris* which presents a profound puzzle (Lockhart et al., 2017). This new strain which is multi-drug resistant has emerged as a major cause of mortality and is posing a serious challenge for health officials the world over (Chowdhary, Sharma, & Meis, 2017; Cortegiani et al., 2018; Jeffery-Smith et al., 2018; Lockhart et al., 2017). While *Candida auris* was reported for the first time in Japan in 2009 it appears to have been *isolated more or less simultaneously* in many widely separated locations across the world.

Phylogenetic analysis by Lockhart et al. (2017) has identified four distinct clades separated by tens of thousands of single nucleotide polymorphisms (SNPs) each of which is geographically localized. A large number of SNPs have been discovered in isolates that were recovered from four widely separated locations (South Asia, East Asia, South America, and South Africa). Whole genome sequencing of these isolates has revealed an exceedingly low genetic diversity within individual regions even across the largest clade involving some 36 isolates from as wide a field as India and Pakistan. The conclusion by Lockhart et al. (2017) is that *C. auris* must have arisen almost simultaneously in multiple four different global locations. Further, from isolates of *Candida* from four continents Lockhart et al. (2017) did not find *C. auris* before 2009 confirming that this pathogen was not simply misidentified previously. While there have been claims that earlier isolates of Candida species may also have been *Candida auris* which were incorrectly identified, these assertions have not been confirmed. Thus, it seems reasonable to conclude that a 2009 date for the origin *Candida auris* is fairly secure (Cortegiani et al., 2018).

Since global cross-infection over a short timescale (<1 year) appears very unlikely one possibility is of independent multiple origins of *Candida auris* from some widely present *Candida* ancestor. A fungicide driver model has been advanced to explain the phenomenon. However this vague model does not fit the available data. We thus argue here that a panspermic in-fall model should be considered as a plausible and better alternative.

Thus, in our explanation (from all the available data) it could be concluded that *C. auris* first arose in 2009 from several environmentally-induced hypermutation events that occurred after in-fall from cosmic (cometary) dust clouds through which the Earth had traversed sometime during or before 2009. Thus this new *C. auris* would appear simultaneously in many widely separated places on the Earth. Alternatively, a genetic hybridization event may have taken place at this time involving a globally distributed set of comet-borne gene segments that were themselves genetically diverse.

How could this have occurred? We critically evaluate the data from a genetic point of view. The data demands that there are at least four pre-existing clades ($\geq$10,000 SNP differences) in an external non-terrestrial source (cometary dust tails) and these came down separately in separate regions and thereafter spread clonally (Lockhart et al., 2017). The other alternative is to consider the existence of a single "mother" or "parent" *C. auris* clade in the cometary dust source, which upon landing and infection of susceptible hosts is induced into a hypermutation-adaptation sequence via a fast, essentially Lamarckian, Adaptive Mutation strategy (Rosenberg, 2001; Chapter "The Efficient Lamarckian Spread of Life in the Cosmos" by Steele et al.) thereby generating in excess of 10,000 new SNP differences from the parent orbiting cosmic strain. The final step that can be envisaged is the dispersal of a successful adaptive variant in a particular region to other hospitals in that region. Thus on the basis of a Panspermic model there are two possible explanations for the strange and striking *C. auris* patterns of genome diversity. The Lamarckian hypermutation strategy at each separate in-fall location (susceptible hospital patients) from a pure line "mother" strain is, on parsimony grounds, preferred.

We have previously argued that a sudden emergence of new pathogenic variants of circulating viruses could be linked to cosmic events related to the well-known 11-year sunspot cycle (Qu & Wickramasinghe, 2017, 2018; Wickramasinghe et al., 2017, 2019). The Earth's magnetosphere, and the interplanetary magnetic field in its vicinity, are both modulated by the solar wind that controls the flow of charged particles onto the Earth. During times of sunspot minima, a general weakening of magnetic field occurs and this would be accompanied by an increase in the flux of galactic cosmic rays (GCR's) and also of charged interstellar and interplanetary dust particles. Evidence for such periodic increases linked to solar activity is evident in the high frequency of noctilucent clouds (as at present, in 2019–2020) and also in the increase of particulate deposits in polar ice cores. Since the latter could, in our view, include biological entities such as bacteria, viruses and other eukaryotic microorganisms like *C. auris*, an increase in their incidence on the Earth will therefore be expected at such times. It is interesting to note that in 2008–2009 (the solar minimum under discussion) the interplanetary magnetic field was the lowest on record since the beginning of the space age. We would therefore expect a significantly enhanced flux of both cosmic rays as well as electrically charged biological entities at this time, so the arrival of a new clade of *C. auris* from a space source should not be ignored.

A crucial fact relating to the first appearance of *Candia auris* in 2009 is that this time marks not merely a solar minimum but *the lowest minimum* of the sunspot cycle in 100 years (see Figs. 1 and 2). This particular minimum was all the more remarkable because the sun was spotless (devoid of spots) for more than 70% of the time. The opportunity of the transference of both Galactic Cosmic Rays (GCR's) and charged molecular structures (e.g., *C. auris*) thus remained continuously optimal over extended periods. At the present time (February 2020) as we approach a new sunspot minimum the sun continues to be exceedingly "quiet" and the expectations are that we are heading for an even deeper minimum than before. The case for epidemiological vigilance for new microbial and viral pathogens cannot be greater than at the present time.

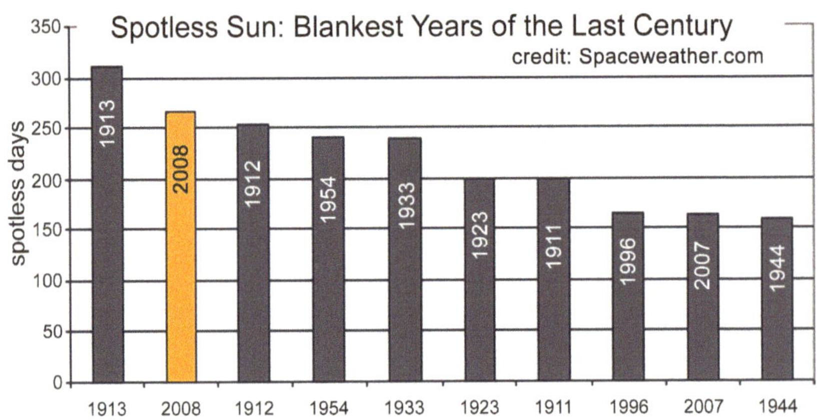

Fig. 1 The distribution of the number spotless days in the Sunspot cycle, showing the period 2008/2009 to be the lowest on record for a century. Exceptionally low sunspot activity in solar minima open the floodgate to interstellar and comet dust.

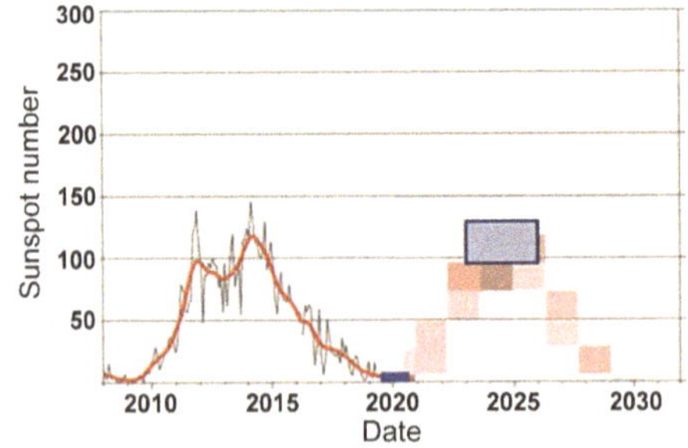

Fig. 2 Current sunspot cycle 24 and predicted cycle 25 (Wickramasinghe et al., 2019). From term solar observations-world data center, Royal observatory of Belgium, Brussels (http:/www.sidc.be/silso/home).

3. Sudden emergence new Coronavirus (COVID-19) causing respiratory infections in Wuhan, China and neighboring regions

We now turn to our critical analysis of the origin of the COVID-19 epidemic underway as we draft this chapter.

3.1 Overview of the COVID-19 epidemic

The global extent of the emotion around this epidemic in the mainstream popular media, and even the scientific press (*Science* magazine) is disturbing. It is without parallel in our experience in this social media internet age. However, it does approach the justified hysteria around the far more serious, and initially more lethal, HIV epidemic/pandemic that suddenly emerged 40 years ago.

The actual COVID-19 viral disease itself causes respiratory "common cold-like" illness in most people diagnosed with symptoms (but many potential carriers of the disease are asymptomatic). The infection can progress to severe pneumonia in elderly and already medically-compromised patients with other conditions (diabetes, coronary disease, etc.). About 2% of all COVD-19 cases have died due to the pneumonia (Fig. 4). Vaccine and antivirals will not help the latter group, but standard well trusted medical care will—to help patients through the respiratory crisis of the life-threatening pneumonia and dangerous inflammatory bronchitis symptoms. Such care will allow recovery of most patients. The fact that "Recoveries" far exceed "Deaths" (Fig. 4) indicates that timely medical care for this otherwise "common cold" respiratory illness must be the medical priority in the epicenter of the infection in Wuhan and nearby regions in China. We believe this medical care is being implemented throughout China.

But it is the origin of this new emergent virus disease which has raised the most angst. It is literally explosively centred on Wuhan, which appears to be the epicenter. And it appeared suddenly without warning. The theory that it jumped from bats via snakes to humans is implausible (below). The same angst over viral origins was also evident when HIV, SARS, MERS, Ebola, and ZIKAV suddenly appeared on the scene. We will not deal with these earlier diseases as their origins, in our considered opinion, are far less clear cut than COVID-19.

However sorting out what is true from what is untrue is a challenge. The current distribution and case numbers as February 14, 2020 are shown in Figs. 3 and 4. The epidemic is centred on the city of Wuhan, in the central Hubei province of China (for an update on our analyses since drafting of this chapter, see Appendix C).

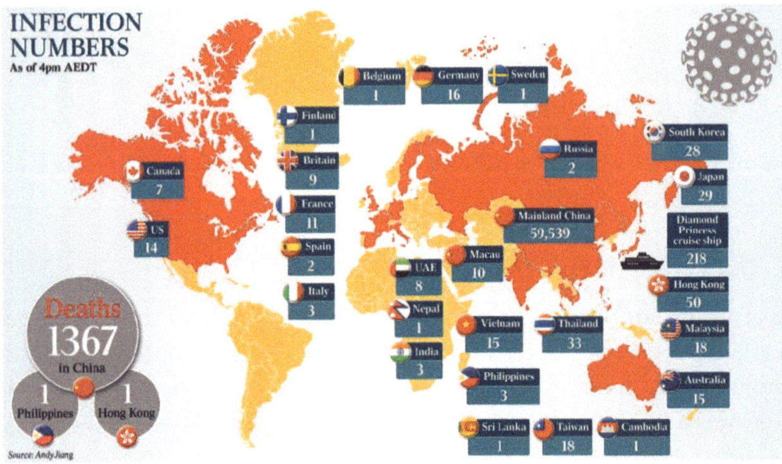

Fig. 3 *The Australian* newspaper as February 14, 2020, p. 8. Illustrative of newspaper reports on the early phase of the COVID-19 pandemic at time of writing in mid-February 2020—infection numbers February 14 2020 are: Canada, 7; US, 14; Finland, 1; Britain, 9; France, 11; Spain, 2; Italy, 3; Belgium, 1; Germany, 16, Sweden, 1; UAE, 8, Nepal, 1; India 3; Macao, 10; Vietnam, 15; Philippines, 3; Sri Lanka, 1; Mainland China, 59,539; Russia, 2; Thailand, 33; Taiwan, 18; Cambodia, 1; South Korea, 28; Japan, 29; Diamond Princes cruise ship, 218; Hong Kong, 50; Malaysia, 18; Australia, 15. Deaths: 1367 China, 1 Philippines, 1 Hong Kong. An update on the total toll of the pandemic, now approaching a global total of 9 million confirmed cases as June 22, 2020 is in Appendix C.

Fig. 4 COVID-19 by numbers, in the Australian newspaper February 14, 2010, p. 8. Figures as of February 14, 6.30 a.m. Source—World Health Organization, Australian Government, AAP, Reuters, other media.

From about mid-January the Chinese government ordered the complete quarantine and lock-down of Wuhan and wider region around the city in Hubei province, affecting 50–100 million people. *ABC News* in Australia estimates Coronavirus (COVID-19) has affected 500 million people in China under lock-down (Updated February 15, 2020, 1:29 a.m.). A problem with all these reports is the lack of detailed information that led officials to such an extraordinary quarantine decision. We speculate later on this.

At the time of writing, the case incidence of this newly discovered Coronavirus is passing through 60,000 and >99.99% of all cases, almost exclusively, are Chinese. From reports of cases that exited Wuhan by aircraft in late January to other countries, say to Australia, the disease does not spread in a sustained way easily person-to-person. But there are clearly apparent cases of person-to-person spread elsewhere (say in United Kingdom and Europe, Box 1). But there is no doubt this disease is centred on China. The Johns Hopkins University COVID-19 case density maps are extremely informative. These are in Figs. 5–7.

To put one interpretation on the striking case patterns in Figs. 5–7, particularly the symmetrical pattern in Fig. 7 it actually looks like a huge viral bomb explosion took place near or over Wuhan and then the radial fall-out of the disease causing viral particles to land on the millions of people either laterally or from above—some of those infected would be susceptible and who then have succumbed to the respiratory illness (in Appendix A, in relation to the expected fall of viruses through the stratosphere is an analysis by way of quantitative analogy, of the expectation of radioactive fall-out patterns from an atmospheric nuclear test in 1958).

Moreover, and paradoxically, asymptomatic patients can be efficient "spreaders" of the disease. This is contrary to normal expectations as usually the infected potential spreader would display overt and full blown disease (and the coughed-up aerosols from such a patient would be dense with viral particles). Indeed, there are wide reports in the media that incubation times can range from 1 to 28 days. But once a potentially infective virus successfully navigates the Innate Immune response (see table 1, Chapter "The Efficient Lamarckian Spread of Life in the Cosmos" by Steele et al.) it would be expected to rapidly multiply within cells and spread peaking in virion numbers 5–7 days later. The actual size of the infective dose is also an important variable. So these wide-range estimates of incubation time reflect, in our view, the variable depth and extent of the *actual physical viral contamination* in the immediate environment of a susceptible subject viz. potentially all objects in the family home environment as well as cars, bicycles, and on their bodies—hair, clothes—and other personal effects, clothes, money, keys, etc. Indeed, the external

BOX 1 Summary of United Kingdom cases (N.C. Wickramasinghe email report to Edward J. Steele).

- January 21, 2020

CDC confirms 1 case of transmission between person who returned from Wuhan, and a person who shared accommodation in the United States.

Via a ski resort in the French Alpine town of Les Contamines-Montjoie, near Switzerland, late last month,

- January 28, 2020

A cluster linked to an Alpine chalet

A British man (Mr. Walsh) from Brighton, was found to have the virus when he returned to the UK (London Gatwick Airport) from Geneva on January 28 on an EasyJet flight. A total of six people in Britain, including Mr. Walsh, and Britons in France who have the virus have been staying in two apartments in a ski chalet in the Alpine resort area near Mont Blanc when they were visited by Mr. Walsh on January 24 *who had attended a* business *conference at the Grand Hyatt Hotel Singapore, where he Is believed to have contracted the virus.*

Mr. Walsh is *thought* to have passed the virus onto eleven confirmed cases while he was at the Ski resort. But he is thought to have come into contact with scores of people after leaving Singapore and no others have yet succumbed.

All the supposed transmissions of the virus from Mr. Walsh to the others were while they occupied the Chalets in France.

Four are from Brighton and Hove. They are Dr. Greenwood and three men, one of whom is a healthcare worker. He also passed it to one other person in the United Kingdom, one person who is now in Mallorca and five United Kingdom nationals in France—one of which is Dr. Greenwood's husband Bob Saynor and another their 9-year-old son. None are said to be in a serious condition.

So far, the places in Brighton and Hove being quarantined are:

- County Oak Medical Centre, where Dr. Catriona Greenwood worked one admin day last week, and its branch surgery at Deneway.
- Grenadier Pub in Hangleton, which was visited by Steve Walsh on February 1.
- Cornerstone Community Centre, where a yoga teacher came into contact with Steve Walsh on February 3. No other people have been advised to self-isolate.
- Easyjet flight EZS8481 to Gatwick from Geneva on January 28, which is believed to be the flight Mr. Walsh took back to the United Kingdom.
- Bevendean Primary School, where a staff member has been in close contact with someone who has been advised to self-isolate (but is not themselves diagnosed).
- Portslade Academy, which told parents on Friday one of its pupils has been advised to self-isolate for a fortnight after coming into contact with the Hove father. It's believed pupils at other schools have been given the same advice.
- Patcham Nursing Home, which has closed its doors to all visitors after being visited by one of the medics now confirmed as having the virus.

The cluster associated with Mr. Walsh could have been coinfected from a common source, with Mr. Walsh showing symptoms first.

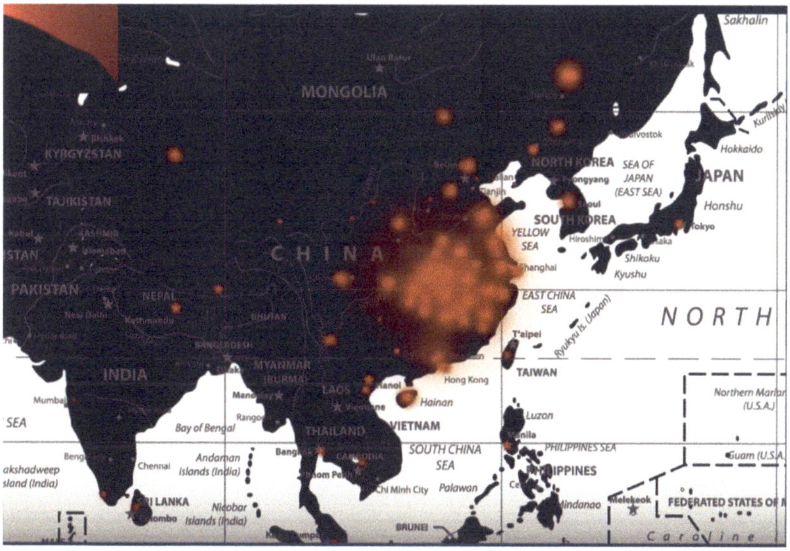

Fig. 5 Case density map—South East Asia region wide. Johns Hopkins University as February 7, 2020. Johns Hopkins University's Centre for Systems Science and Engineering.

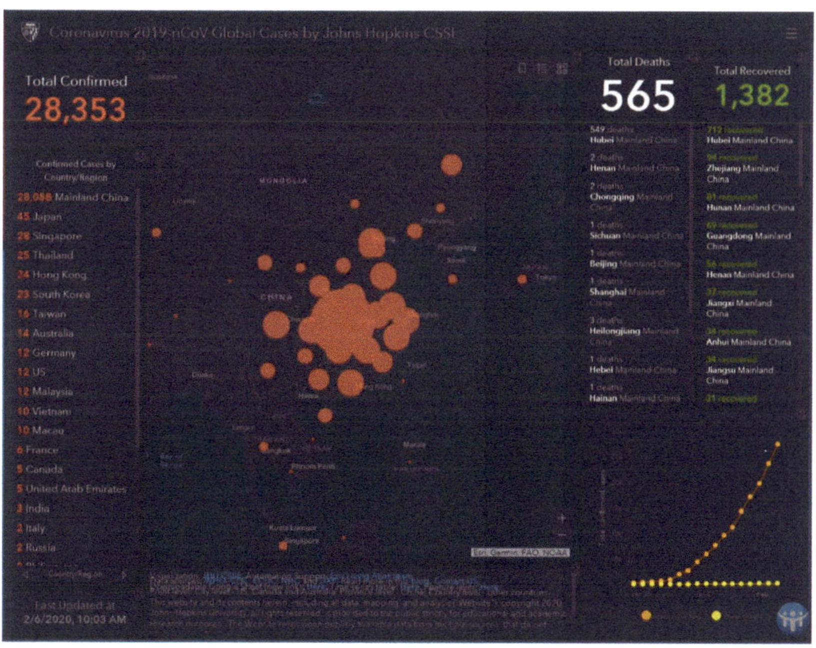

Fig. 6 Case density map—China and nearest neighbors. Johns Hopkins University as February 7, 2020. Johns Hopkins University's Centre for Systems Science and Engineering.

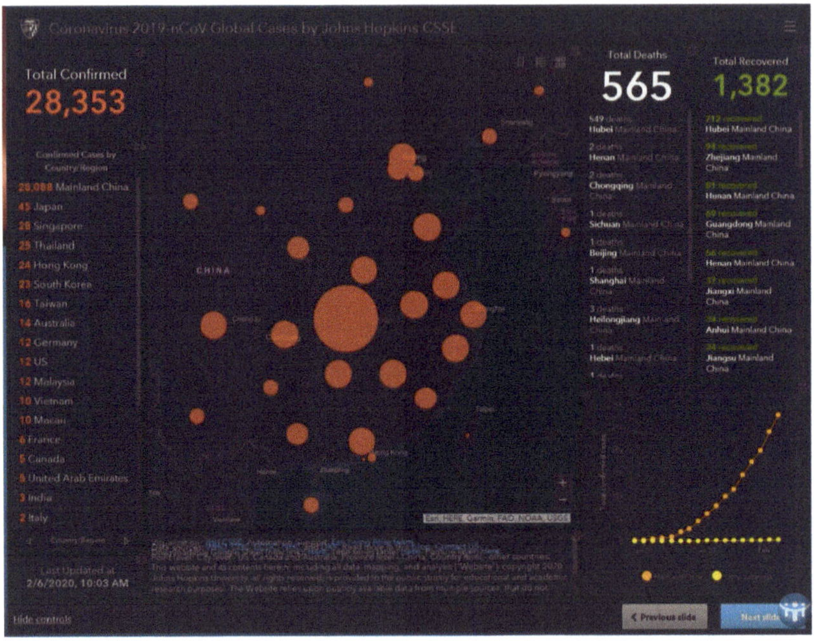

Fig. 7 Case density map—China itself. Johns Hopkins University as February 7, 2020. Johns Hopkins University's Centre for Systems Science and Engineering.

surface of the face mask may be the main carrier of the physical contamination. We explore the extent of viral environmental contamination further below.

3.2 Detailed analysis of COVID-19 epidemic

We now analyze all reliable genetic, epidemiological and geophysical and astrophysical data. This leads to the alternate hypothesis that COVID-19 arrived via a meteorite, a presumed relatively fragile and loose carbonaceous meteorite, that struck North East China on October 11, 2019. This is at odds with the main stream expert "Infectious Disease" opinion of traditional person-to-person spread of an infectious endemic disease such as, for example, Cholera (*Vibrio cholerae*).

We then assume the viral debris and particles then made land fall in the Wuhan and related regions about a month to 6 weeks later resulting in first cases of the viral pneumonia caused by COVID-19 emerging in Wuhan regions late November 2019-early December 2019 (Cohen, 2020; Huang et al., 2020). Such an hypothesis is indeed consistent with the striking patterns shown in Fig. 7. And it makes, therefore, an extraordinary set of predictions, that we will explore at some length in our conclusions.

It suggests, first, massive region-wide physical contamination with potentially trillions of infective COVID-19 viral particles in central China-contaminating buildings, roadways, cars and factory equipment, vegetation, surface water pools, people (and their clothes, body parts such as hair, skin, personal affects, mobile phone, keys, wallets, etc.) as well as wild and domestic animals, etc. This would explain the actions of the Chinese Government who are acting to appear to be in receipt of such presumed knowledge (or information) from region-wide sampling to detect COVID-19 RNA sequences in swabs from physical objects, people and animals (via real time PCR).

The recent paper by Huang et al. (2020) and the extremely important news commentary by Cohen in *Science* (Cohen, 2020) highlights many unusual aspects of the outbreak of COVID-19. The evidence demonstrates that many cases of disease (about 30% of case reports) arose in locations unconnected with the Wuhan seafood and meat market, and the total tally continues to increase. Phylogenetic analyses of COVID-19 (previously named nCov-2019) sequences show little by way of sequence variation thus indicating low mutation rates thus approximating closely to what would be expected for a pure culture, of a single infecting and replicating sequence affecting disease cases (Andersen, 2020; Lu et al., 2020). These facts are now combined with the global epidemiological data, that points in the main to little or no really sustained human-to-human transmission thus far (e.g., latest reports by the Australian Department of Health). We are aware there are apparent exceptions, e.g., the "super-spreader" from Singapore, via the French Alps, and then to a United Kingdom GP surgery reporting mild symptoms, resulting in the GPs also getting the disease (Box 1). We interpret that spread by viral contamination of physical objects in the main rather than direct "cough in your face" human to human spread.

In any case, current data suggest that the human-to-human spread rate is unusually low, and may be dependent on proximity and dose of virus delivered at very close quarters. The "lethality" or "death rate" from this or any other epidemic disease increases in older patients with pre-existing conditions so wider global estimates yield a death rate at 2% of infected (Fig. 4). All these basic facts now appear agreed.

The initial traditional explanation of the new epidemic of COVID-19 is that it jumped from bats (possibly via snakes) to humans and then spread by human-to-human infection contact mutating at a high rate. This explanation is at odds with the data at present. Indeed Jon Cohen the respected *Science* magazine journalist reports that the head of the Huang et al. (2020) study when interviewed said:

> *Bin Cao of Capital Medical University, the corresponding author of* The Lancet *article and a pulmonary specialist, wrote in an email to* ScienceInsider *that he and his co-authors "appreciate the criticism" from Lucey (Daniel Lucey, an infectious disease specialist at Georgetown University confirmed the epidemic could* not possibly be caused by visits to the Wuham seafood and meat market).
>
> *"Now It seems clear that [the] seafood market is not the only origin of the virus," he wrote. "*But to be honest, we still do not know where the virus came from now." *(our italics)*

Indeed Dr. Bin Cao speaks for all mainstream medical and epidemiological professionals around the world—no formal traditional explanation can be provided for the origins of COVID-19. Thus Andrew Rambaut, Professor of Molecular Evolution at the University of Edinburgh tweeted: "Don't think any epidemiologist is still thinking that a non-human animal reservoir has had anything to do with the nCoV-2019 epidemic since December.

Certainly the genome data doesn't support that." (Reported in Heidi Han and Kieran Gair, Associated Press, *The Australian* newspaper January 27, 2020.)

Thus, when we combine all the available facts we cannot rule out a viral in-fall event targeting the Wuhan province and the wider region around it as an explanation as a first cause of the epidemic. This would fit with the admittedly heterodox view of viral pandemics first proposed by Hoyle and Wickramasinghe as far back as 1978 (Hoyle & Wickramasinghe, 1979, 1990; Wickramasinghe et al., 2019; Wickramasinghe, Wainwright, & Narlikar, 2003). This concept accords with the theory of cosmic biology for which growing evidence have recently been presented in the Chapters of this book and in recent peer-reviewed papers where all the main extant evidence has been reviewed and is consistent with the Hoyle-Wickramsinghe thesis (Steele et al., 2018; Steele, Al-Mufti, et al., 2019; Steele, Gorczynski, et al., 2019). Our theory thus posits a sporadic input of cosmic bacteria, viruses and other micro-organisms that has the potential to interact with evolving terrestrial life forms, causing terrestrial diseases and further adaptive evolution on Earth.

3.3 Link with a direct strike of meteorite over central-North East China, October 11, 2019

In the case of the current coronavirus epidemic in China it is interesting to note that an exceptionally bright fireball event was seen on October 11, 2019 over Sonjyan City in the Jilin Province of NE China (see Fig. 8). It is tempting to speculate that this event (although it happened hundreds of kilometers distant from Hubei) had a crucial role to play in what is now unfolding in and

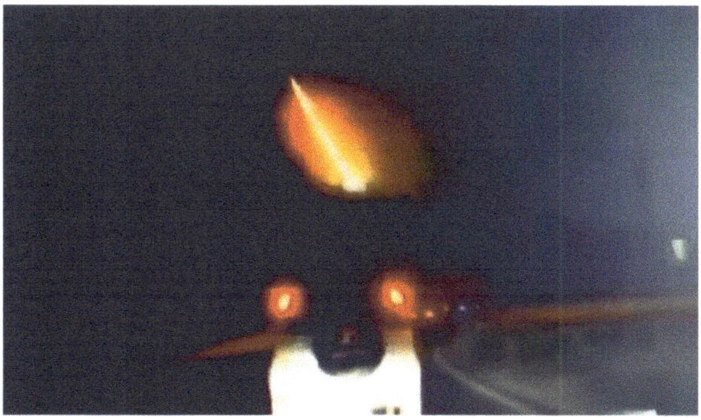

Fig. 8 The public record of this meteorite strike can be found at the Space.com website in an article by Tariq Malik, on October 13, 2019 "Brilliant Midnight Fireball Lights Up Sky Over Northeast China". The October event is described at: https://www.space.com/china-midnight-meteor-brilliant-fireball-october-2019.html.

throughout China. Indeed, the match with the Johns Hopkins University case incidence patterns is so striking it is difficult to easily dismiss this as a chance correspondence of patterns, in both time and place (e.g., Fig. 7).

If a fragment of a fragile and loosely held carbonaceous meteorite carrying a cargo of trillions of viruses/bacteria and other primary source cells (for the cosmic replication of the COVID-19 virus), may have entered the mesosphere and stratosphere at high speed ~30 km/s, its outer, loosely-held envelope carrying a biological cargo may have got dispersed in the mesosphere stratosphere and troposphere. Indeed, a much larger original meteoroid could easily have been fragmenting and dispersing its contents before the ignition of the fireball event. A reasonable assumption is that the fireball which struck 2000 km N of Wuhan may have been part of a wide tube of debris the bulk of which was deposited in the stratosphere to fall over Wuhan. The fall time through the atmosphere of 1–10 μm-sized solid particles could range from a few months to well over a year on the basis of straightforward calculations (e.g., in the Appendix of Hoyle & Wickramasinghe, 1979 "Diseases from space"). Because dispersal at ground level depends on the vagaries of meteorology and precipitation the deposition of virus at ground level is expected to be patchy in regard to both time and place. This is certainly consistent (thus far) with what has happened in relation to the new COVID-19 coronavirus epidemic between November 2019 and the present day (February 15, 2020). Following the initial deposition of infective particles in a small localized region (e.g., Wuhan, Hubei province, China) particles that have already become dispersed over a wider area in the troposphere will fall to ground in a higgledy-piggledy manner, and this process could be extended over a typical timescale of 1–2 years until an initial inoculant of the infective agent would be drained. This accords well with many new strains of viruses including influenza that have appeared in recent years (Wickramasinghe et al., 2019).

The possible link of sunspots with pandemics has been discussed over many years (Qu & Wickramasinghe, 2017, 2018; Wickramasinghe et al., 2017, 2019) and is worthy of brief further discussion. The present cycle (interface between cycles 24 and 25, Fig. 2) has seen the lowest minimum for well over a century with many sunspot free days recorded in the last months of 2019. Sunspot minima are associated with a weakening of the interplanetary magnetic field near the Earth, which in turn allows easy ingress of Galactic Cosmic Rays (GCRs) and electrically charged bacteria and viruses to the Earth. The mutagenic role of GCRs can cause genetic changes in already circulating viruses, but it is primarily to an enhanced flux of new infective particles released by the exploding meteoroid that we turn. Indeed a perfect storm over China is paying out before our eyes, a meteorite delivering COVID-19 particles corresponding with a very significant Sunspot Minimum cycle. It raises the important issue: How would other densely populated countries have reacted to, and handled, this event involving COVID-19? It was only the vagaries of chance that it exploded over China.

4. Conclusions

We conclude by noting some predictions and expectations:

- We expect the pattern of further spread of the new Coronavirus (COVID-19) to be dictated mostly by primary in-fall until a high level of person-to-person infectivity might *possibly* be achieved and the virus then acquires the status of an endemic virus.
- Viral contamination of the "environment" in the most general sense explains most of the apparent contagion, e.g., news reports like in Box 1.
- Thus, the possibility cannot be ruled out that the *Diamond Princess* cruise ship (and the more recent Westerdam cruiseship) in the South China Sea was contaminated by a fragment of the main COVID-19 dust cloud. Similar inexplicable events appeared to happen for ships at sea during the 1918–1919 Spanish Flu Pandemic (Hoyle & Wickramasinghe, 1979).
- And, further to this, other drifting COVID-19 smaller dust clouds that have not as yet made land fall may target remote island and other communities, as was also the case during the 1918–1919 Spanish Flu Pandemic (Hoyle & Wickramasinghe, 1979).

- Given the low mutation rate, the very wide apparent in-fall infectivity pattern (Fig. 7) the expectation is this pure viral culture has inoculated millions of Chinese citizens (as well as potentially millions of wild and domestic animals in China) inducing protective adaptive immune responses (Acquired Herd Immunity) on a very large scale.
- Thus, development of a so called "COVID-19 vaccine" which is much in the news at the time of writing would be a waste of public tax-payer funds if mounted on the scale envisaged by governments and national centers for disease control.
- We thus expect the decline of the epidemic (peaking and declining at time of writing) to be driven by this mass natural vaccination process now underway in China. So, the suddenly emerging COVID-19 epidemic, like many similar suddenly emerging human epidemics in the past (SARs, MERs, ZIKAV), is expected to rapidly end by the self-limiting processes of wide spread herd immunity (a pattern likely to be repeated in other countries, Appendix C).
- We thus expect that the incidence of serum antibodies specific for COVID-19 to be wide-spread in the Chinese population in the coming months. So, millions will be potentially immunized for life against future infections with COVID-19.
- How long will COVID-19 remain potentially infective in the physical environment? Clearly for some time—given that over the space of a month or so many cases appeared rapidly, spread by environmental contamination in our view, and not by traditional person-to-person generated aerosols at the height of the donor's infection. This is consistent with those news reports out of China "As the death toll rose to 80, China said, increasing concerns about the potential the virus was infectious even before symptoms were visible rapidity of its spread." (Heidi Han and Kieran Gair, Associated Press, *The Australian* newspaper January 27, 2020.)

5. Postscript

As this chapter was submitted to the publisher an authoritative news despatch from Japan reports sporadic outbreaks across the country with *no direct link* with China (Appendix B). Further, in early February we tried to alert the world on our interpretation of the origins of COVID-19 with many of the same arguments and analyses listed in this chapter. One succinct letter was sent to *The Lancet*, and the other was a more general article for a wider lay readership, to *The Australian* newspaper—both articles were rejected by the editors. The archived PDFs of both articles can be found at the viXra.

org site under accession numbers URLs http://viXra.org/abs/2002.0039?ref=11076818, and https://vixra.org/abs/2002.0118. However an expanded comments on the origin and spread of the 2019 Coronavirus (COVID-19) has now been published in Wickramasinghe et al. (2020a, 2020b, 2020c), and see Steele and Lindley (2020) also in discussion in Appendix C.

Acknowledgment

We thank Professor Sanjaya Senanayake for bringing the *Candida auris* data to our attention and for discussions.

Appendix A

On the fall of viruses through the stratosphere

The defining feature of infectious diseases which are caused by biological entities that arrive from space relates to the manner in which they come to be distributed over the Earth's surface. If such microbial agents are introduced via the agency of cometary bolides that survive frictional heating in the mesosphere, their fragmentation and dispersal as clouds of particles in the stratosphere will determine the way in which they finally arrive at the surface of the Earth.

The falling speed of spherical particles of various sizes (with a notional density 1 g cm^{-3}) through the atmosphere can be calculated as a function of height (y-axis) was calculated from formulae and data given by Kasten (1968). The results are shown in Fig. A1. We note from the extreme right curve of Fig. A1 that a particle of radius 10 μm falls through the lowest 10 km of the stratosphere at a speed of ~1 cm/s and so takes only a few days to cover what for smaller particles is the slowest part of their downward journey. All such particles fall comparatively rapidly through the mesosphere (z = 80–50 km), and then more and more slowly down through the stratosphere below z = 50 km. A particle with the size of a typical bacterium ~1 μm, falls through the lowest 10 km of the stratosphere at a speed of about 2×10^{-2} cm/s and thus falls to ground in a time-scale of $\sim 5 \times 10^{7}$ *s*, that is about 2 years. Because there is more of the stratosphere through which such a particle must fall at high latitudes than in the tropics (the troposphere being higher in the tropics) the slow part of the journey is more extended the higher the latitude. A bacterium falling in ~1 year in the tropics would fall in ~2 years in temperate latitudes and in 2–3 years towards the poles.

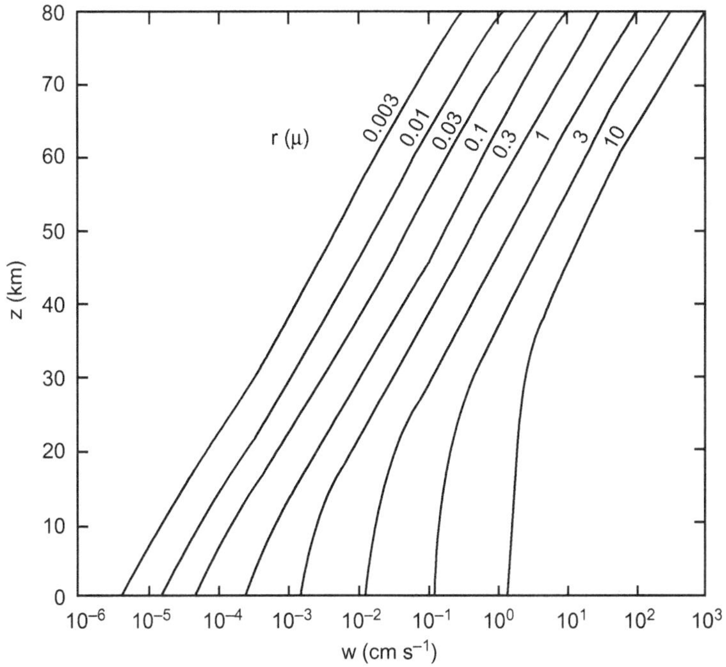

Fig. A1 Falling speed of spherical dust particles as a function of height in the atmosphere. *Reproduced from Hoyle & Wickramasinghe. 1982,* Proofs that life is cosmic, *Mem. Inst. Fund. Studies, Sri Lanka. Copyright Chandra Wickramasinghe.*

If a particle of the size of a typical virus, a particle say with a diameter of ~0.03 μm, fell under gravity through still air the timescale for the slowest part of the journey through the bottom 10 km of the stratosphere (z = 20–10 km) will be $\sim 10^9$ s, that is about 30 years (see Fig. A1). This is so slow that other means of descent involving large-scale air movements in the stratosphere have to be considered. Further, the possibility of large clumps of viruses encased in cosmic dust particles will also change the relative effective sizes of infective particles and consequently their speeds of entry. We say this because … "The sophistication of viral infectivity and their *modus operandi* of cell-cell spreading does not end here … viral genomes … can be propagated almost indefinitely (Combe, Garijo, Geller, Cuevas, & Sanjuan, 2015) as 'virion clusters' of mixtures of infective and crippled genomes with significant numbers of newly minted virus particles enwrapped in protective membrane vesicles—a type of multiunit nanoparticle. This then constitutes the actual infective dose (*or unit*) rather than just a single exported virion entering a nearby target cell to cause a productive infection as is commonly believed (Chen et al., 2015; Combe et al., 2015)." Text and references from Steele et al. (2018).

Thus, although in general vertical mass movements of air are feeble compared to those in the troposphere, some vertical stratospheric movement takes place despite the inhibiting effect of an inverted temperature gradient. The physical mechanism for mass stratospheric movements is the equator-to-pole temperature difference which acts as a heat engine across parallels of latitude, a heat engine that operates more strongly the larger the temperature difference—i.e., much more strongly in winter than in the summer. A similar mechanism applies also to the troposphere, where an engine crossing parallels of latitude transfers heat from tropical regions towards the poles, again more in winter than in the summer.

Ozone measurements can be used to trace the mass movements of air in the stratosphere. Such measurements show a winter downdraft that is strongest over the latitude range 40–60 degrees. Taking advantage of this annual downdraft, individual virus particles (or dust-encased clumps) incident on the atmosphere from space would therefore reach ground-level generally in temperate latitudes, which therefore emerge from these considerations as the regions of the Earth where upper respiratory infections are likely to be most prevalent. This is true on the supposition that the Earth is smooth, which of course it is not. The exceptionally high mountains of the Himalayas on the North India-China boundary rearing up through most of the height range to the stratosphere, introduce a large perturbation to the smooth condition, which may be expected to affect adversely this particular region of the Earth, particularly China. In effect the Himalayan peaks, higher than 8 km, could act as a drain plug for most of the viruses incident on the atmosphere at latitude about 30 degrees N. The vast 1.4 billion population of China will thus be inundated by this drainage effect making China the quickest and worst affected region of the planet for cosmic pathogen-particle in-falls. As a consequence other parts of the Earth at about 30 degrees N could fortunately be largely free of incoming viral particles, unless it happens that such particles are incident as components within larger particles.

A direct proof that the winter downdraft effect in the stratosphere occurs overwhelmingly over the latitude range 40–60 degrees was demonstrated by Kalkstein (1961). In the last of the series of nuclear bombs that were tested in the atmosphere, a radioactive tracer element Rh-102 was introduced into the atmosphere at a height above 100 km and the fall out of the tracer was measured month by month through airplane and balloon sampling at altitudes of ~20 km. The radioactive tracer Rh-102 took more than a decade to clear itself through repeated seasonal downdrafts of the kind we

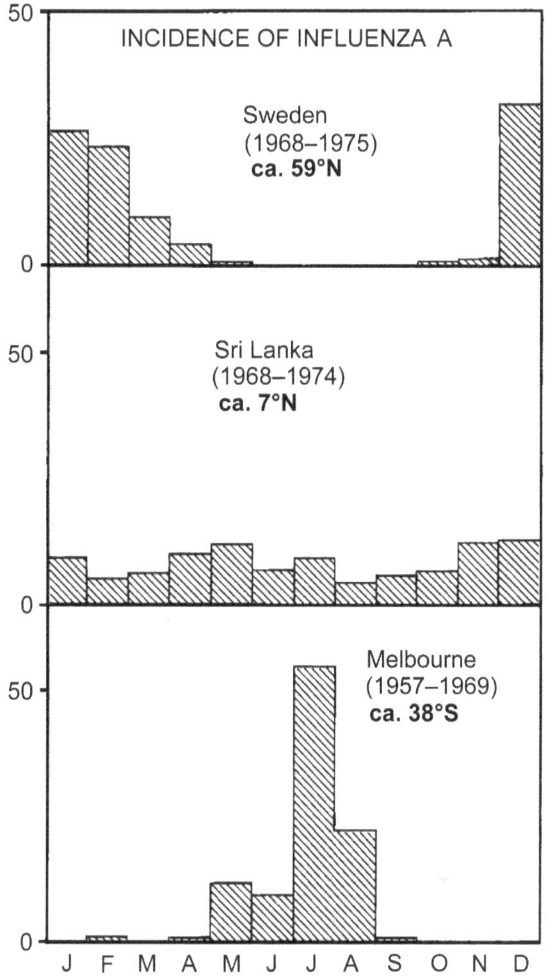

Fig. B1 Incidence of seasonal influenza A in three countries. *Reproduced from Hoyle & Wickramasinghe. (1982).* Proofs that life is cosmic. *Figure copyright Chandra Wickramasinghe.*

have described. The fall out was found to be much greater at temperate latitudes than elsewhere with the period January to March being the dominant months for the Northern Hemisphere. This is exactly similar to the pattern of incidence of influenza and other seasonal respiratory viral diseases in Northern temperate latitudes, a situation that is well-known to every medical practitioner and health authority. In the Southern hemisphere the situation is similar but 6 months displaced. This is clearly evident in Fig. B1. Note that in a tropical location such as Sri Lanka no discernible seasonal effect can be detected in the data between 1968 and 1974.

Appendix B as February 15, 2020

Headline: None of Japan's new coronavirus patients had direct China links

First death raises fear that virus is quietly spreading

By Yusuke Kurabe, Nikkei staff writer, in Nikkei Asian Revie

February 13, 2020 22:37 JST Updated on February 14, 2020 04:52 JST

https://asia.nikkei.com/Spotlight/Coronavirus/None-of-Japan-s-new-coronavirus-patients-had-direct-China-links

- A Kanagawa Prefecture woman in her 80s died from the coronavirus. Her son-in-law also tested positive for the disease. A doctor in Wakayama Prefecture and a man in Chiba Prefecture are confirmed to have the virus. None of them traveled to China recently or had contact with people who visited Hubei Province, the epicenter of the outbreak. The 80 year old woman's symptoms began January 22 when she felt fatigue, the health ministry said. Symptoms worsened on January 25, prompting her to see a doctor 3 days later. She was placed under observation. The victim was hospitalized February 1, diagnosed with pneumonia. She underwent screening for the coronavirus Wednesday. The test results came back positive Thursday, the day she died. Her son-in-law also tested positive for the coronavirus. The man, a taxi driver in his 70s living in Tokyo, has been hospitalized since February 6, but the symptoms are reportedly mild. He developed a fever January 29.
- A doctor in Wakayama Prefecture south of Osaka has been infected with the virus, prefectural officials said Thursday. The man, in his 50s, has been hospitalized with symptoms of pneumonia, but is otherwise in stable condition. The doctor did not travel outside the country in the 14 days prior to the onset of symptoms, nor can any contact with people coming from China be confirmed. Wakayama officials suspect the infection had domestic origins.
- Elsewhere, a man in his 20s from Chiba Prefecture near Tokyo is also confirmed to have the virus. He developed a fever and other symptoms February 2. The man reportedly has not travelled overseas or had contact with other infected individuals.
- Besides the outbreak on the Diamond Princess cruise ship, which has infected over 200 people aboard the vessel quarantined in Yokohama, 29 cases of coronavirus had been confirmed inside Japan through Wednesday.

- These cases raise new challenges for health officials, who until now had been trying to contain the virus by closely monitoring people with the possibility of contracting the disease. If more people with no direct links to China become sick, determining infection routes will become impossible. Instead of containment, treating seriously sick people may have to become the priority.

Appendix C

An update on published and submitted work by the group as June 22, 2020

The COVID-19 pandemic has now engulfed almost all parts of the globe inhabited by human beings. Confirmed cases are approaching 9 million and confirmed global deaths as June 22 are 468,589. We have updated our analysis here as the pandemic further unfolded. These analyses are now in several papers which have been published (or under-submission) since this paper was drafted. The most important developments are covered in the new citations in the reference list (and below), particularly our discussion on the progress of the major explosive outbreaks following the original epidemic in Wuhan through December-January 2020. Initially, in February, this appeared to go from China across the Pacific to the US West Coast (Wickramasinghe et al., 2020b), but it then became apparent the putative viral-laden meteorite dust clouds moved along the north 40° latitude band heading in a westerly direction from China towards Europe (Wickramasinghe et al., 2020c). We speculate this transportation took place mainly in the Stratosphere and upper Troposphere, and this helps explain in part the major explosive outbreaks of COVID-19 in the temporal order, Tehran/Qom, Italy/Lombardy, Spain then New York City all on the north 40° latitude band. The genetic sequence of the virus has been analyzed where available in each of the major explosive locations, particularly Wuhan and New York City (Steele & Lindley, 2020). Contrary to a widely held perception that the COVID-19 is a hypermutating contagious virus it is still the case, at the time of writing, that the most vulnerable at risk of death through respiratory complications are the co-morbid elderly. Further, the virus does not hypermutate, it appears to a use a riboswitch-mediated haplotype variation strategy which could be associated with the ethnic-genetic background of the subjects infected. Thus the genetic sequence of the virus in Wuhan/China, West Coast USA (February, 2020), Spain and New York City is essentially $\geq$99.98% identical apart from a small set of key variant putative riboswitch sites (ranging from 2 to 5 on average across the length of the 29,903 nt RNA genome) which define a given haplotype. Thus, there is much haplotype homogeneity in China, but considerable haplotype diversity in New York City. The virus appears to adapt to new host genetic backgrounds by switching putative haplotype presumably allowing superior RNA replication in that cellular environment. The only factor we cannot explain then is why the super explosions of the viral epidemics in certain locations and not others. One likely factor would be the infective dose of the virus which we predict was far greater at these epicentre locations because it came in as a high dose in-fall in meteorite dust from the upper atmosphere.

References

Andersen, K. (2020). *Clock and TMRCA based on 27 genomes Novel 2019 coronavirus*. http://virological.org/t/clock-and-tmrca-based-on-27-genomes/347.

Bradley, S. F. (2019). Candida auris infection. *JAMA*, *322*, 1526.

Chatterjee, S., Alampalli, S. V., Nageshan, R. K., Chettiar, S. T., Joshi, S., & Tatu, U. S. (2015). Draft genome of a commonly misdiagnosed multidrug resistant pathogen Candida auris. *BMC Genomics*, *16*, 686.

Chen, Y.-H., Du, W.-L., Hagemeijer, M. C., Takvorian, P. M., Pau, C., et al. (2015). Phosphatidylserine vesicles enable efficient en bloc transmission of enteroviruses. *Cell*, *160*, 619–630. https://doi.org/10.1016/j.cell.2015.01.032.

Chow, N. A., Gade, L., Tsay, S. V., Forsberg, K., Greenko, J. A., Southwick, K. L., et al. (2018). Multiple introductions and subsequent transmission of multidrug-resistant Candida auris in the USA: A molecular epidemiological survey. *Lancet Infectious Diseases*, *18*, 1377–1384.

Chowdhary, A., Sharma, C., & Meis, J. F. (2017). Candida auris: A rapidly emerging cause of hospital-acquired multidrug-resistant fungal infections globally. *PLoS Pathogens*, *13*(5), e1006290. https://doi.org/10.1371/journal.ppat.1006290.

Cohen, J. (2020). Wuhan seafood market may not be source of novel virus spreading globally. *Science*, January 26. https://doi.org/10.1126/science.abb0611.

Combe, M., Garijo, R., Geller, R., Cuevas, J. M., & Sanjuan, R. (2015). Single-cell analysis of RNA virus infection identifies multiple genetically diverse viral genomes within single infectious units. *Cell Host Microbe*, *18*, 424–432. https://doi.org/10.1016/j.chom.2015.09.009.

Cortegiani, A., Misseri, G., Fasciana, T., Giammanco, A., Giarratano, A., & Chowdhary, A. (2018). Epidemiology, clinical characteristics, resistance, and treatment of infections by Candida auris. *Journal of Intensive Care*, *6*, 69. https://doi.org/10.1186/s40560-018-0342-4.

Forsberg, K., Woodworth, K., Walters, M., Berkow, E. L., Jackson, B., Chiller, T., et al. (2018). Candida auris: The recent emergence of a multidrug-resistant fungal pathogen. *Medical Mycology*, *57*, 1–12.

Hoyle, F., & Wickramasinghe, N. C. (1979). *Diseases from space*. London: J.M. Dent & Sons, Ltd.

Hoyle, F., & Wickramasinghe, N. C. (1990). Influenza—Evidence against contagion: Discussion paper. *Journal of the Royal Society of Medicine*, *83*(4), 258–261.

Huang, C., Wang, Y., Li, X., Ren, L., Zhao, J., Hu, Y., et al. (2020). Clinical features of patients infected with 2019 novel coronavirus in Wuhan, China. *The Lancet*, *395*, 497–506. Published online January 24,2020. https://doi.org/10.1016/S0140-6736(20)30183-5.

Jackson, B. R., Chow, N., Forsberg, K., Litvintseva, A. P., Lockhart, S. R., & Welsh, R. (2019). On the origins of a species: What might explain the rise of Candida auris? *Journal of Fungi*, *5*, 58.

Jeffery-Smith, A., Taori, S. K., Schelenz, S., Jeffery, K., Johnson, E. M., Borman, A., et al. (2018). Candida auris: A Review of the Literature. *Clinical Microbiology Reviews*, *31*(1), pii: e00029-17. https://doi.org/10.1128/CMR.00029-17.

Kalkstein, M. I. (1961). Rhodium-102 high altitude tracer experiment. *Science*, *137*, 645.

Kasten, F. (1968). Falling speed of aerosol particles. *Journal of Applied Meteorology*, 7, 944.

Lockhart, S. R., Etienne, K. A., Vallabhaneni, S., Farooqi, J., Chowdhary, A., Govender, N. P., et al. (2017). Simultaneous emergence of multidrug-resistant Candida auris on 3 continents confirmed by whole-genome sequencing and epidemiological analyses. *Clinical Infectious Diseases*, *64*, 134–140. https://doi.org/10.1093/cid/ciw691.

Lu, R., Zhao, X., Li, J., Niu, P., Yang, B., Wu, H., et al. (2020). Genomic characterisation and epidemiology of 2019 novel coronavirus: Implications for virus origins and receptor binding. *The Lancet*, *395*, 565–574. published Online January 29, 2020. https://doi.org/10.1016/S0140-6736(20)30251-8.

Pfaller, M. A., Diekema, D. J., Turnidge, J. T., Castanheira, M., & Jones, R. N. (2019). Twenty years of the SENTRY antifungal surveillance program: Results for Candida species from 1997–2016. *Open Forum Infectious Diseases*, *6*(S1), S79–S94.

Qu, J., & Wickramasinghe, C. (2017). SARS, MERS and the sunspot cycle. *Current science*, *113*(8), 1501–1502.

Qu, J., & Wickramasinghe, C. (2018). Weakened magnetic field, cosmic rays and the Zikavirus outbreak. *Current science*, *115*(3), 382–383.

Rhodes, J., & Fisher, M. C. (2019). Global epidemiology of emerging Candida auris. *Current Opinion in Microbiology*, *52*, 84–89.

Rosenberg, S. M. (2001). Evolving responsively: Adaptive mutation. *Nature Reviews Genetics*, *2*, 805–815.

Steele, E. J., Al-Mufti, S., Augustyn, K. A., Chandrajith, R., Coghlan, J. P., Coulson, S. G., et al. (2018). Cause of Cambrian explosion—Terrestrial or cosmic? *Progress in Biophysics and Molecular Biology*, *136*, 3–23. https://doi.org/10.1016/j.pbiomolbio.2018.03.004.

Steele, E. J., Al-Mufti, S., Augustyn, K. A., Chandrajith, R., Coghlan, J. P., Coulson, S. G., et al. (2019). Cause of Cambrian explosion—Terrestrial or cosmic?—Reply to commentary by R Duggleby. *Progress in Biophysics and Molecular Biology*, *141*, 74–78.

Steele, E. J., Gorczynski, R. M., Lindley, R. A., Liu, Y., Temple, R., Tokoro, G., et al. (2019). Lamarck and Panspermia—On the efficient spread of living systems throughout the cosmos. *Progress in Biophysics and Molecular Biology*, *149*, 10–32. pii: S0079-6107(19)30112-9. https://doi.org/10.1016/j.pbiomolbio.2019.08.010.

Steele, E. J., & Lindley, R. A. (2020). Analysis of APOBEC and ADAR deaminase-driven Riboswitch Haplotypes in COVID-19 RNA strain variants and the implications for vaccine design. Accepted for publication in *Research Reports*, June 22, 2020, under Quantitative Biology. https://vixra.org/abs/2006.0011?ref=11360015.

Wickramasinghe, N. C. (2012). DNA sequencing and predictions of the cosmic theory of life. *Astrophysics and Space Science*, *343*, 1–5.

Wickramasinghe, N. C., & Steele, E. J. (2016). Dangers of adhering to an obsolete paradigm: Could Zika virus lead to a reversal of human evolution? *Journal of Astrobiology & Outreach*, *4*(1). https://doi.org/10.4172/2332-2519.1000147.

Wickramasinghe, N. C., Steele, E. J., Gorczynski, R. M., Temple, R., Tokoro, G., Qu, J., et al. (2020a). Comments on the origin and spread of the 2019 Coronavirus. *Virology: Current Research*, *4*(1). https://doi.org/10.37421/Virol Curr Res.2020.4.109.

Wickramasinghe, N. C., Steele, E. J., Gorczynski, R. M., Temple, R., Tokoro, G., Wallis, D. H., et al. (2020b). Growing evidence against global infection-driven by person-to-person transfer of COVID-19. *Virology: Current Research*, *4*(1). https://doi.org/10.37421/Virol Curr Res.2020.4.110.

Wickramasinghe, N. C., Steele, E. J., Gorczynski, R. M., Temple, R., Tokoro, G., Kondakov, A., et al. (2020c). Predicting the future trajectory of COVID-19. *Virology: Current Research*, *4*(1). https://doi.org/10.37421/Virol Curr Res.2020.4.111.

Wickramasinghe, N. C., Steele, E. J., Wainwright, M., Tokoro, G., Fernando, M., & Qu, J. (2017). Sunspot cycle minima and pandemics: The case for vigilance. *Journal Astrobiology and Outreach*, *5*, 2. https://doi.org/10.4172/2332-2519.1000159.

Wickramasinghe, C., Wainwright, M., & Narlikar, J. (2003). SARS—A clue to its origins? *The Lancet*, *361*, 1832.

Wickramasinghe, N. C., Wickramsinghe, D. T., Senananyake, J., Qu, J., Tokoros, G., Temple, R., et al. (2019). Space weather and pandemic warnings? *Current Science*, *117*(10), 1554 (25 Nov 2019).

CHAPTER SEVEN

A cosmic virosphere

Predrag Slijepcevic[a] and N. Chandra Wickramasinghe[b,c,d,*]
[a]Brunel University, London, United Kingdom
[b]University of Buckingham, Buckingham, United Kingdom
[c]Centre for Astrobiology, University of Ruhuna, Matara, Sri Lanka
[d]National Institute of Fundamental Studies, Kandy, Sri Lanka
[*]Corresponding author: e-mail address: ncwick@gmail.com

Contents

Abstract

The concept of a cosmic virosphere that serves as the repository of information for all life on Earth and throughout the Universe is discussed. Recent studies in geology, astronomy and biology point to an intimate connection between the evolution of life and a cosmic virosphere/biosphere.

1. Introduction

"Have humans evolved beyond nature—and do we even need it?" asked a recent article on a popular portal (https://theconversation.com/have-humans-evolved-beyond-nature-and-do-we-even-need-it-128790). The real-life answer arrived few weeks later, in the form of the global Covid-19 pandemic. We have not evolved beyond nature and never will. In the immortal words of Lynn Margulis "*We can no more be cured of our viruses than we can be relieved of our brains' frontal lobes: we are our viruses*" (Margulis, 1998, p. 82).

Advances in Genetics, Volume 106
ISSN 0065-2660
https://doi.org/10.1016/bs.adgen.2020.06.001

Lynn Margulis was acutely aware of the fact that we are built from viruses. Around 45% of the human genome consists of sequences that belonged to viruses (Moelling & Broecker, 2019). Margulis promoted the idea of symbiogenesis. She thought that life is more a collaborative game between equal partners—viruses, bacteria, plants, and animals—and less the ruthless competition between cheating predators. And she was right in many respects. In the hierarchy of nature, a virus is no less important than a human being.

2. The rationale for a cosmic virosphere

It has recently been recognized that the total biomass of our planet may be dominated by viruses. Epifluorescence microscopy of sea water has recently turned up with a total count of $0.5–1 \times 10^{10}$ virions per liter or $\sim 10^{31}$ virions throughout the oceans. It seems most likely that these viruses in surface sea water are being constantly resupplied through deposits from the atmosphere. The total flux of viruses falling through the atmosphere and reaching the 4 km peaks of the Sierra Nevada mountains in Spain was recently measured by Reche, D'Orta, Mladenov, et al. (2018). The average flux of viruses falling to this height was found to be *800 million per square meter per day*. While some of this flux could be of Earth origin it is by no means clear that all of it could be explained in this way. Moreover, earlier sampling of microbiota in the stratosphere at heights of 41 km yielded estimates of bacterial infall of some *20–200 million bacteria per square meter per day plus 10 times as many viruses,* and this is from a height where lofting from the ground is considered unlikely. There is a tendency to dismiss these results as necessarily due to contamination from the Earth itself, but this is largely based on an inherent disbelief that the Earth is connected to a wider cosmic biosphere and virosphere.

Viruses have partners. The most common viruses on Earth are those that infect bacteria (Moelling & Broecker, 2019). Bacteria are the second most frequent biogenic structures on Earth. The bacteriosphere and virosphere are thus structurally coupled (Slijepcevic, 2018, 2019). They have been existing in partnership for the last 4 billion years here on Earth, and possibly much longer throughout the cosmos.

3. A dynamic microbiome—Evidence of co-evolution

The recent recognition of a microbiome associated with evolved lifeforms including humans emphasizes the intimacy of the partnership between microorganisms and higher life forms that continue to the present day. The human gut microbiota has often been described as the most densely populated ecosystem on Earth (Penders, Stobberingh, Savelkoul, & Wolffs, 2013; Turnbaugh et al., 2009). Included here are many groups of extremophiles, including acidophiles, that are significantly different from free-living microbial populations. Most importantly they have the proven ability to transfer viral particles (virions) to invading bacteria.

4. Endogenous viruses

In discussing the possibility of either the influenza virus or the COVID-19 virus is of cosmic origin the one line disproof that is often offered is that a virus that had originated in an extraterrestrial environment cannot have any possibility of interacting with evolved life forms on Earth (Wickramasinghe, Steele, Gorczynski, et al., 2020). This argument ceases to be valid if the evolution of terrestrial life involved the interaction with these same viruses over long evolutionary timescales.

The role of viruses and of a virosphere interfacing the evolution of terrestrial life first became evident after whole genomes, including the human genome, came to be sequenced in 2001. One of the surprises that followed was the discovery that what were once thought to be "junk DNA" actually had a viral origin. We thought that some 45% of our DNA consist of retroviral sequences (endogenous retroviruses, ERVs)—RNA viruses that have reverse transcribed their RNA into DNA (Moelling & Broecker, 2019). Their significance in causing disease as well as in contributing to evolution is only just beginning to be understood.

New evidence also points to past episodes of retroviral infections (of which HIV is an example) in almost all mammalian species. De Groot, Otting, Doxiadis, et al. (2002) have identified an entire repertoire of genes (MHC class 1 genes) in chimpanzees that confer immunity against chimpanzee-derived simian immune deficiency virus. The extant chimp populations thus appear to be descendants of a small cohort of animals that had actually survived a lethal HIV-like pandemic in the distant past.

In this and other examples the survivors of past pandemics that carry ERVs and HERVs in their DNA bear testimony to a history of prior infections. Recent studies have also shown that endogenization of viruses is not confined to retroviruses. For instance, an RNA sequence of non-retroviral origin has been found incorporated in the germ line of several mammalian species, including rodents around 40 million years ago (Horie, Honda, Suzuki, et al., 2010). With a significant fraction of the human genome comprised of viral sequences, the logical conclusion is that the evolution of hominids leading to *Homo sapiens* involved a series of viral pandemics followed by endogenization of their gene sequences into the evolving germ line. The question as to how viruses from an external cosmic virosphere appear to be "fortuitously" fit to interact with terrestrially evolved lifeforms like ourselves then becomes irrelevant. Viruses of cosmic origin are already in our genomes and have contributed to our evolution over long periods of time.

5. First signs of life on Earth

The cosmic life paradigm leads to testable predictions (Hoyle & Wickramasinghe, 2000). Recent discoveries have shown that the earliest evidence of microbial evidence of life on the Earth is preserved within zirconium crystals within rocks in the Allans Hill formation in Western Australia that were laid down 4.2 billion years ago during the Hadean geological epoch (Bell, Boehnke, Harrison, & Mao, 2015). At this time the Earth was bombarded with comets and asteroids and the case for asserting that the first bacterial life came with comets is now beginning to look undeniable.

6. Evolution of cosmic life

The theory of cometary panspermia posits that bacteria and viruses continue to arrive at the Earth from their first moment of arrival to the present day (Hoyle & Wickramasinghe, 1979, 1981, 2000). The evolution of terrestrial life throughout this period took place against the background of new viruses and bacteria bringing new genes to augment and direct evolutionary progress. Comets are the primary repositories, incubators and transporters of bacteria and viruses across the Universe.

7. Mixing of genes on a galactic scale

While comets could supply a source of primitive life (archeae, bacteria, viruses) to interstellar dust clouds and thence to planets like Earth (which we now know probably exist in their billions in the galaxy alone) genetic products of evolved life could also be disseminated on a galaxy-wide scale. Our present day solar system which is surrounded by an extended halo of some 100 billion comets (the Oort Cloud) moves around the center of the galaxy with a period of 240 My. Every 40 million years, on the average, this cometary cloud is perturbed due to the close passage of a molecular cloud. Gravitational interaction leads to hundreds of comets from the Oort Cloud being injected into the inner planetary system, some to collide with the Earth. Such collisions can not only cause extinctions of species (as one impact surely did 65 million years ago, killing the dinosaurs), but they could also result in the expulsion of surface material containing terrestrial biomaterial back into space.

A mechanism thus exists for the genes of evolved Earth-life to be transferred to alien habitable planets. A fraction of the Earth-debris so expelled survives shock-heating and could be laden with viable microbial ecologies. Such life-bearing terrestrial material could reach newly forming planetary systems in the passing molecular cloud within a million years of the ejection event. A new planetary system thus comes to be infected with terrestrial microbes and terrestrial genes that can contribute, via horizontal gene transfer, to an ongoing process of local biological evolution. Once life begins and evolves on an alien planet in the new system, the same process can be repeated transferring genetic material carrying local evolutionary "experience" to other molecular clouds and planetary systems.

If every life-bearing planet can transfer genes in this way to more than one other planetary system (say 1.1 on the average) with a characteristic time of 40 My then the number of seeded planets after 9 billion years (age of the Galaxy) is $(1.1)^{9000/40} \approx 2 \times 10^9$. Such a large number of "infected" planets illustrates that evolution, involving horizontal gene transfers, must operate not only on the Earth or within the confines of the solar system but must extend over a truly galactic scale. Life throughout the galaxy on this picture would constitute a single connected biosphere. The cosmic biosphere is made up of two distinct but intimately interacting components—bacteria and viruses.

8. Distinction between bacteria and viruses

There is a crucial distinction between these two forms of biological entities. Bacteria are independent entities that are in a strict sense alive; viruses are not. Biologists think that a living entity must synthesize proteins and replicate autonomously. Viruses do neither. They are classified as parasites that require bacteria, plants or animals to live. However, the term parasite may be the consequence of our anthropocentric bias. How can a creative system, such as nature, be dominated by parasites? The answer of course is that they are by no means parasites. They form the information repository for the whole of cosmic biological evolution.

Viruses could be conceived as invisible fingers on the hand of nature, that has been sculpting organisms through scattering viral RNA and DNA, from the dawn of life. That's why our genomes, and the genomes of all plants and animals, are full of viral sequences (Margulis, 1998). Collectively viruses contain the information content for the emergence and evolution of all life in the universe—all life that has been and ever shall be manifest in the cosmos.

The idea of nature as the sculptor, must not be seen as the evidence of intelligent design—the creator separate from its creation, on the contrary. Nature the sculptor, is the natural mind at work—the nature making itself. This idea was articulated and developed by the British cyberneticist and philosopher, Gregory Bateson. He called the democracy of the biosphere—where all organisms have equal rights—the ecology of mind (Bateson, 2000). For Lynn Margulis and James Lovelock, the ecology of mind was equivalent to the homeostasis of the Earth biosphere or Gaia (Lovelock & Margulis, 1974). But Gaia may only be a part of the enormous galactic biosphere. Was Anaxagoras right when he speculated about the cosmic mind or *Nous*?

Nature is more of an artist, than a scientist. Viruses present in us, and all other living beings, are traces of the nature's artistic hand. As Heidegger (1977) argued, the greatest sculpture ever is nature itself. In the epic act of sculpting itself, a process biologists call autopoiesis, nature cannot, in a truly artistic manner, predict its next state. We humans are just a temporary thought in the mind of nature.

Yet we are easily deceived by the idea of self-deification, which we see as our natural right. According to Zimmerman (2008), the idea of human self-deification persists in the European civilization from St Paul and Martin Luther to Hegel and the modern day trans-humanists. The real-life burial of the self-deification idea is viruses and bacteria that have been infecting people every day since our emergence on this planet. The viral and bacterial infections are traces of the nature's invisible hand, no more powerfully as is now manifest in the COVID-19 pandemic.

References

Bateson, G. (2000). *Steps to an ecology of mind.* Chicago: University of Chicago Press.

Bell, E. A., Boehnke, P., Harrison, T. M., & Mao, W. L. (2015). Potentially biogenic carbon preserved in a 4.1 billion-year-old zircon. *PNAS, 112*, 14518–114521.

De Groot, N. G., Otting, N., Doxiadis, G. G. M., et al. (2002). Evidence for an ancient selective sweep in the MHC class I gene repertoire of chimpanzees. *PNAS, 99*, 11748–11753.
Heidegger, M. (1977). *The question concerning technology and other essays. Translated by William Lovitt.* New York & London: P Garland Publishing Inc.
Horie, M., Honda, T., Suzuki, Y., et al. (2010). Endogenous non-retroviral RNA virus elements in mammalian genomes. *Nature, 463*, 84–87.
Hoyle, F., & Wickramasinghe, N. C. (1979). *Diseases from space.* London: J.M. Dent & Sons.
Hoyle, F., & Wickramasinghe, N. C. (1981). *Evolution from space.* London: J.M. Dent & Sons.
Hoyle, F., & Wickramasinghe, N. C. (2000). *Astronomical origins of life.* Dordrecht: Kluwer Academic Publishers.
Lovelock, J. E., & Margulis, L. (1974). Atmospheric homeostasis by and for the biosphere: The Gaia hypothesis. *Tellus, 26*, 2–10.
Margulis, L. (1998). *The symbiotic planet.* London: Phoenix.
Moelling, K., & Broecker, F. (2019). Viruses and evolution—Viruses first? a personal perspective. *Frontiers in Microbiology, 10*, 523.
Penders, J., Stobberingh, E. E., Savelkoul, P. H., & Wolffs, P. F. (2013). The human microbiome as a reservoir of antimicrobial resistance. *Frontiers in Microbiology, 4*, 87.
Reche, I., D'Orta, G., Mladenov, N., et al. (2018). Deposition rates of viruses and bacteria above the atmospheric boundary layer deposition rates of viruses and bacteria above the atmospheric boundary layer. *The ISME Journal, 12*(4), 1154–1162. https://doi.org/10.1038/s41396-017-0042-4.
Slijepcevic, P. (2018). Evolutionary epistemology: Reviewing and reviving with new data the research programme for distributed biological intelligence. *Biosystems, 163*, 23–35.
Slijepcevic, P. (2019). Principles of information processing and natural learning in biological systems. *Journal for General Philosophy of Science* (in press). https://doi.org/10.1007/s10838-019-09471-9.
Turnbaugh, P. J., Hamady, M., Yatsunenko, T., Brandi, L. C., Alexis, D., et al. (2009). A core gut microbiome in obese and lean twins. *Nature, 457*, 480–484.
Wickramasinghe, N. C., Steele, E. J., Gorczynski, R. M., et al. (2020). On the fragility of empires and paradigms. *Virology: Current Research, 4* (in press). http://doi.org/10.37421/Virol Curr Res.2020.4.112.
Zimmerman, M. E. (2008). The singularity: a crucial phase in divine self-actualization? *Cosmos and History: The Journal of Natural and Social Philosophy, 4*, 347–370.

CHAPTER EIGHT

An internet of microbes straddling the cosmos

Predrag Slijepcevic[a] and N. Chandra Wickramasinghe[b,c,d,*]
[a]Brunel University, London, United Kingdom
[b]University of Buckingham, Buckingham, United Kingdom
[c]Centre for Astrobiology, University of Ruhuna, Matara, Sri Lanka
[d]National Institute of Fundamental Studies, Kandy, Sri Lanka
*Corresponding author: e-mail address: ncwick@gmail.com

Contents

Abstract

Exchanges of information analogous to a global internet have been known to take place between biological systems on the Earth ranging from bacteria and viruses to plants and animals. We argue that this process can be extended to include a cosmic biosphere within which evolution would seem to be intimately interlinked across astronomical, perhaps cosmological distance scales. Comets and interstellar dust, argued to have a bacterial/viral component, could be involved in establishing these links.

1. Introduction

In a 1981 book with the title "*Space Travellers: The bringers of life*" Fred Hoyle and one of us (Hoyle & Wickramasinghe, 1981a) wrote as follows:

> *The potential of bacteria to increase vastly in their number is enormous. It should occasion no surprise, therefore, that bacteria are widespread throughout astronomy. Rather would it be astonishing if biological evolution had been achieved on the Earth alone, without the explosive consequences of such a miracle ever being permitted to emerge into the Universe at large. How could the Universe ever be protected from such a devastating development? This indeed would be a double miracle, first of origin, and second of terrestrial containment.*

Advances in Genetics, Volume 106
ISSN 0065-2660
https://doi.org/10.1016/bs.adgen.2020.06.002

Some biologists have probably found themselves in opposition to our arguments for the proprietary reason that it seemed as if an attempt were being made to swallow up biology into astronomy. Their ranks may now be joined by those astronomers who see from these last developments that a more realistic threat is to swallow up astronomy into biology...

Over the 4 decades that followed the merging of these two disciplines has certainly taken place, and the parallels and interactions between the two have become obvious at every turn. Intelligence that was once considered the prerogative of humans has been co-opted in machines up to a point of artificial intelligence being considered an inevitable adjunct to biological life.

The internet of things is a burgeoning scientific and technological concept, implying the development of new modes of communication between individuals, devices, and the external world. Although it is conceived as a synthesis between high speed internet connections that are already in place and human consciousness, it can also be seen as an extension of biologically-driven connections that already exist. We shall discuss this development in the context of the universe at large which can be seen as an extension of the interactions that are found to exist between viruses and bacteria on the Earth.

2. Bacteria, viruses and comets

Data has accumulated over several decades on how starlight is dimmed as it passes through interstellar dust clouds as shown in Fig. 1.

From an analysis of this data Fred Hoyle and one of the present authors came to startling conclusion nearly 40 years ago (see articles in Hoyle & Wickramasinghe, 2000). The wavelength dependence of scattering and absorption of starlight is consistent with a surprisingly simple model. A mixture of bacteria, nanobacteria and viruses produce a remarkable fit to all the data without making any other assumption than that the particles in interstellar clouds responsible for scattering and extinction are closely similar to the distribution of microbiota on the Earth today. The correspondences are illustrated in Fig. 2, but with their optical properties modified only by the vacuum conditions that exist in space.

Basically two types of solid particles appear to dominate the dust in interstellar clouds. One is characterized by freeze-dried bacteria with an average radius of 0.3 μm and the other by viruses/viroids/nanobacteria of average radius 0.01 μm. The arguments supporting this conclusion have not in way diminished since they were first proposed; indeed they have

Fig. 1 Horsehead Nebula.

only increased to the point of being close to compelling (Steele, Gorczynski, Lindley, et al., 2019). The only criticism to be voiced thus far is that this would require the assumption that Earth-like microbiota are present cosmically, which is generally thought to be impossible. The correspondences between astronomical data and this model are shown in Fig. 1 implies the following.

The average mass densities of bacteria and viruses in interstellar space turn out to be as follows:

$$\text{Mass density of bacteria } 10^{-26}\ \text{g/cm}^3;\ \text{Mass density of viruses } 10^{-27}\ \text{g/cm}^3.$$

This in turn implies that a significant mass fraction of the carbon in the universe is tied up in a cosmic biological system—a cosmic biosphere. The average distances between viruses in interstellar space can be easily estimated as $\sim$10 m and between bacteria about 300 m. These distances are possibly not close enough for direct communication, particularly from the deep-freeze state. There is however a possibility (perhaps remote) that some form of radio signaling might be possible because elongated bacteria (bacilli) in space are electrically charged to a potential of a few volts, and due to collisions

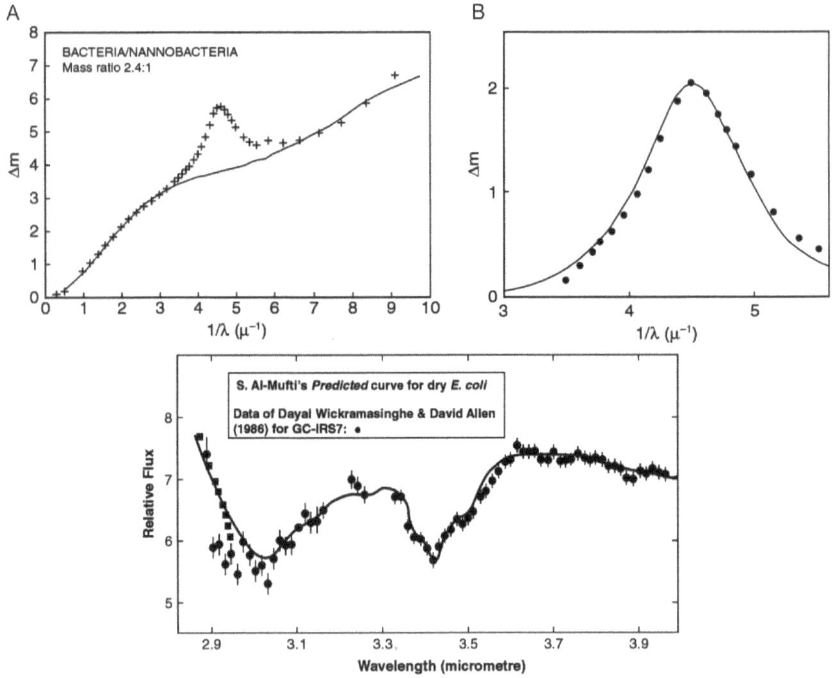

Fig. 2 (Upper A) The mean extinction curve of the galaxy (points) compared with the contribution of desiccated bacteria and nanobacteria. (Upper B) The residual extinction compared with the normalized absorption coefficient of an ensemble of 115 biological aromatic molecules. (*Lower*) The first detailed observations of the Galactic center infrared source GC-IRS7 compared with earlier laboratory spectral data for dehydrated bacteria. See reference articles in Hoyle and Wickramasinghe (2000).

with gas atoms, they can be shown to rotate at radio frequencies (Hoyle & Wickramasinghe, 1970). However, interactions may happen within the radioactively heated liquid domains of comets which according to the theory of cometary panspermia (Hoyle & Wickramasinghe, 1981b) are the long-term repositories of cosmic bacteria and viruses. The symbiosis between viruses and bacteria that is clearly evident on the Earth (Slijepcevic & Wickramasinghe, 2020) may thus be seen to have an origin in comets.

The main reservoir of comets in our solar system is a shell (Oort Cloud) of some 100 billion individual comets. Within the Oort cloud (or similar clouds of comets in exoplanetary systems) micrometeorite impacts could lead to transfers of microbiota via sputtering and/or evaporation of life-laden material, and in this way some interaction/intercommunication between cometary bodies can become established. However, a longer much range of communication will follow from the steady loss of comets from a planetary system such as our solar system to interstellar space, and thence to other planetary systems.

The theory of cometary panspermia implies that life on Earth came from a cometary reservoir of primitive life systems—bacteria, viruses in particular. These basic cellular/genetic units come to be incorporated in comets which serve as repositories, incubators, and distributors of biology throughout the Universe (Hoyle & Wickramasinghe, 2000). Radioactive heat sources maintain warm liquid water domains within comets which serve as culture media of primitive life. Evidence of the first bacterial life delivered by comets to the Earth have recently been found in the form of microbial fossils that date back to 4.2 billion years ago, a time when we know the Earth suffered an episode of severe bombardment by comets. The first life to Earth was delivered in this way, and subsequent interaction with comets provided the genetic material that contributed to the evolutionary history of our planet.

NASA's orbiting Kepler telescope since its launch in 2009 has discovered nearly 4000 definite as well as probable habitable exoplanets by a sampling only a relatively small volume of the Milky Way in our vicinity. Most of these exoplanets orbit red dwarf stars that are on the average twice the age of our sun. Extrapolating from the present sample the estimated total number of habitable planets in the entire Milky Way is reckoned to be in excess of 100 billion. These planetary systems are almost certainly endowed with comets, similar to our solar system that is surrounded by a shell of over a 100 billion comets.

Our solar system is estimated to lose of the order of 10 comets per year in hyperbolic orbits. If we assume that each one of the 100 billion similar exoplanetary systems in the galactic disc loses ~10 comets per year, the rate of injection of comets into the interstellar medium is:

$$dN/dt \approx 10^{11} \times 10 = 10^{12}\ \mathrm{yr}^{-1}.$$

and in a turn-over time (star formation timescale) of the ISM of $\sim 3 \times 10^9$ years the steady state value of N is

$$N \cong 10^{12} \times 3 \times 10^9 = 3 \times 10^{21}.$$

The number density of interstellar comets is therefore estimated to be

$$\rho \approx 3 \times 10^{21}/10^2 = 3 \times 10^{19}\,\mathrm{kpc}^{-3}$$

taking $\sim 10^2 \, kpc^3$ to be the volume of the galactic disc. The mean separation between interstellar comets will be $\sim \rho^{-1/3}$–10^{15} cm, a distance close to the average orbital distance of Neptune. These distances, although large, still permit biological connections via material transfers through cometary activity (Biver, Rauer, Despois, et al., 1996). The frozen outer crust could rupture due to internal biological activity in radioactively heated sub-surface domains (Wickramasinghe, Hoyle, & Lloyd, 1996) causing high pressure gas high speed jets to escape. Ejecta may have speeds of 300 km/s so that physical transfers between interstellar comets could take place over a timescale of $\sim$1 year. Adjacent interstellar comets could well be interlinked in a communication network involving iterant microbes in this way.

Within individual comets themselves the possibility of a network of liquid water "veins" maintained by metabolic heat and radioactivity is a possibility that has recently been brought to light in relation to new data relating to microbes in Antarctic ice (Junge, Eicken, & Deming, 2004; Parrilli, Sannino, Marino, et al., 2011).

3. Microbial network on the Earth

One of us (PS) has recently argued, based on the existing microbial and higher life-forms data, that the entire Earth biosphere is connected through the myriad of information networks (Slijepcevic, 2018, 2019). For example, bacteria have a sophisticated communication system that fulfills the classical linguistic criteria applicable to the human communication system. Similarly, all higher organisms can communicate through different forms of semiosis. Furthermore, there is an effective communication between organisms that belong to different species groups. Bacteria living in our intestines, communicate with our brain cells, a system called the gut-brain axis. This is a form of cross-kingdom communication based on biosemiotics.

The entire communication system of the Earth biosphere has been named the interactome (Slijepcevic, 2019). The components of the interactome are bacteria and archea, protists, fungi, plants, animals, and ecological units such as biomes integrating all organisms. We can now add viruses to this giant network. For example, the evidence is emerging that viruses have the communication capacity. They gain this capacity as soon as they infect cells and communicate through the chemical that has been named "arbitrium" (Slijepcevic, 2019). The biosphere thus becomes a supersystem of interacting biological components. Importantly, the biosphere as an information system is not centralized but it is distributed. As such, it is resilient to any damage that can occur as a result of extra-terrestrial impacts and can quickly recover.

Theoretical models suggest that the system core components are bacteria (Gilbert & Neufeld, 2014). If bacteria are removed from the biosphere, the entire system will gradually decline and cease to exist. This is consistent with the long-known fact that bacteria are the key organisms involved in regulating biogeochemistry. The Earth biosphere existed for at least 2 billion years as the bacteria-dominated system (archea are far less frequent).

However, the model fails to appreciate the importance of viruses for the bacterial dominance on Earth. Since bacteria are alive (capable of synthesizing proteins and replicating autonomously) and viruses are not (do not synthesize proteins; do not replicate autonomously) it seems likely that the viral contribution to the partnership is to act as a reservoir of the biological functions through integrating their genetic sequences into bacteria where they become the source of new proteins. These new proteins serve to enrich the bacterial metabolic ranges, and allow them to survive in the harshest environments. Therefore, it seems reasonable to argue that the viral-bacterial partnership may be essential for maintaining the Earth biosystem. This is partially confirmed recently in a study that generated a new tree of life according to which bacteria dominate the Earth biosphere, with Archea and Eukarya being only minor components (Hug et al., 2016).

4. Synthesis and concluding remarks

When the above biological data are integrated with the arguments outlined in the first part of the article, a new picture emerges that is theoretically compatible with the model implying two key things. First, the forms of life outside the Earth are likely to be predominantly microbial, involving bacterial-viral partnerships. This is consistent with the standard assumption that the bacterial-viral partnership has been existing in continuity on Earth for 4 billion years, which is almost one third of the duration of cosmos. Given that bacteria and viruses are structurally coupled, it is likely that they cannot exist without each other anywhere in the cosmos. Second, it is unlikely that any other organisms, apart from bacteria and viruses, can exist in the interior of comets.

If these two assumptions are accepted, it becomes possible to argue that there must be some form of communication between bacteria and viruses throughout the cosmos. The communications may be intense in permissive environments such as the interior of comets and could continue unabated if the conditions are permissive in the new environments following cometary impacts. For example, when a comet impacts the planetary crust its interior will be released, including the microbes. If the new planet contains basic inorganic chemicals in its crust, microbes will continue to live. Even the most bizarre chemical systems can be explored for subsistence of the bacterial-viral partnership (extremophiles). Through the mutation of the bacterial genetic material and infections of bacteria with mutating viruses, new ways of exploiting even the most bizarre nutrients can be invented. To create a new planet-wide biosphere a critical mass of microbes is required. This suggests that multiple cometary impacts scattered all over the virgin planet may be required to turn it into the living Earth-like planet.

In the conditions of less permissive environments, the communication between bacteria and viruses may cease as bacteria turn into the spore form of existence. This is likely to happen in the interstellar space. Nevertheless, flashes of energy emitted irregularly by nearby stars, in the manner of solar flares, may revive the dormant bacteria and allow viral infections if there are viruses nearby. If re-awakened microbes are in the vicinity of a habitable planet they may simply fall on that planet and continue to live in the new conditions.

In summary, the microbial information network that links the Earth in densely packed information flows, may exist in the extra-terrestrial environments in different densities. More dense networks on the habitable planets with permissive environments and less dense in the inter-planetary spaces. One of us (PS) dubbed the microbial communication network the bacterial World Wide Web (Slijepcevic, 2017), following the elaboration of this concept by Sonea and Mathieu (2001). The bacterial WWW is enriched with the higher life forms that emerged through bacterial-archeal-viral mergers (eukaryogenesis) turning the original web into the biological equivalent of the Internet of Living Things.

Some versions of the bacterial-viral WWW and the Internet of Living Things are likely to exist in the other parts of the Universe, and perhaps even throughout it.

References

Biver, N., Rauer, H., Despois, D., et al. (1996). Substantial outgassing of CO from comet Hale-Bopp at large heliocentric distance. *Nature*, *380*(6570), 137–139.

Gilbert, J. A., & Neufeld, J. D. (2014). Life in a world without microbes. *PLoS Biology*, *12*(12), e1002020. https://doi.org/10.1371/journal.pbio.1002020.

Hoyle, F., & Wickramasinghe, N. C. (1970). Radio waves from grains in HII regions. *Nature, 227*, 473.
Hoyle, F., & Wickramasinghe, C. (1981a). *Space travellers—The bringers of life*. Cardiff: University College.
Hoyle, F., & Wickramasinghe, N. C. (1981b). Comets—a vehicle for panspermia. In C. Ponnamperuma (Ed.), *Comets and the origin of life*. D. Reidel. *Nature*, 1996 Mar 14; 380 (6570): 137–139.
Hoyle, F., & Wickramasinghe, N. C. (2000). *Astronomical origin of life*. D. Reidel.
Hug, L. A., Baker, B. J., Anantharaman, K., Brown, C. T., Probst, A. J., Castelle, C. J., et al. (2016). A new view of the tree of life. *Nature Microbiology, 1*(5), 16048.
Junge, K., Eicken, H., & Deming, J. W. (2004). Bacterial activity at 2 to 20°C in Arctic wintertime Sea Ice. *Applied and Environmental Microbiology, 70*, 550–557.
Parrilli, E., Sannino, F., Marino, G., et al. (2011). Life in icy habitats: New insights supporting panspermia theory. *Rendiconti Lincei Scienze Fisiche e Naturali*. https://doi.org/10.1007/s12210-011-0136-2.
Slijepcevic, P. (2017). *Microbes have their own version of the Internet*. https://theconversation.com/microbes-have-their-own-version-of-the-internet-75642.
Slijepcevic, P. (2018). Evolutionary epistemology: Reviewing and reviving with new data the research programme for distributed biological intelligence. *Biosystems, 163*, 23–35.
Slijepcevic, P. (2019). Principles of information processing and natural learning in biological systems. *Journal for General Philosophy of Science* (in press). https://doi.org/10.1007/s10838-019-09471-9.
Slijepcevic, P. and Wickramasinghe, N.C. 2020. In press.
Sonea, S., & Mathieu, L. G. (2001). Evolution of the genomic systems of prokaryotes and its momentous consequences. *International Microbiology, 4*, 67–71.
Steele, E. J., Gorczynski, R. M., Lindley, R. A., et al. (2019). Lamarck and panspermia—On the efficient spread of living systems throughout the cosmos. *Progress in Biophysics and Molecular Biology, 149*, 10–32. https://doi.org/10.1016/j.pbiomolbio.2019.08.010.
Wickramasinghe, N. C., Hoyle, F., & Lloyd, D. (1996). Eruptions from comet Hale-Bopp at 6.5AU. *Astrophysics and Space Science, 240*, 161–165.

CHAPTER NINE

Is the 2019 novel coronavirus related to a spike of cosmic rays?

N. Chandra Wickramasinghe*
University of Buckingham, Buckingham, United Kingdom
Centre for Astrobiology, University of Ruhuna, Matara, Sri Lanka
General Sir John Kotelawala Defence University, Ratmalana, Sri Lanka
National Institute of Fundamental Studies, Kandy, Sri Lanka
*Corresponding author: e-mail address: ncwick@gmail.com

Contents

Abstract

WHO's pronouncement of the 2019 novel coronavirus outbreak as a pandemic disease came months after we published a warning that the present deepest minimum of the sunspot cycle would be likely to facilitate the onset of a viral pandemic. During a deep sunspot minimum (deepest in 100 years) such as we are now witnessing, two space related phenomena could have an effect on the disposition of viral disease and potential pandemics. With the weakening of the magnetic field in the Earth's vicinity, there would be a high flux of mutagenic cosmic rays. These processes would be likely to herald the onset of new pandemics. Neutron counts from Moscow Neutron Monitor show that the flux of cosmic rays reaching Earth in 2019 was indeed at a maximum over a timespan of half a century since 1962. It is of interest to note that immediately prior to the first recorded cases of the novel Corona virus in China a peak of cosmic rays was measured as is indicated by the Huon neutron monitor data. Recent research revealed that estimates of the timing of the most recent common ancestor of COVID-19 made with current sequence data point to emergence of the virus in late November 2019 to early December 2019, compatible with the earliest retrospectively confirmed cases and the cosmic ray spike in late November 2019. In our view, this strong cosmic ray spike was in some way connected with the onset of the outbreak.

On December 31, 2019, a cluster of pneumonia of unknown etiology was reported in Wuhan City, Hubei Province of China. Although the outbreak of pneumonia was determined to be caused by the 2019 Novel Coronavirus (COVID-19), the true cause of its sudden emergence remains unclear.

Advances in Genetics, Volume 106
ISSN 0065-2660
https://doi.org/10.1016/bs.adgen.2020.06.003

K.G. Andersen, recently wrote in *Nature Medicine* that scientific evidences show that the new coronavirus is the result of natural evolution, not laboratory synthesis (Andersen, Rambaut, Lipkin, et al., 2020). Here, we can scientifically explain this conclusion in further. Generally speaking, there are several ways that are normally considered for the emergence of new viruses. One is that these are viruses that exist in nature but have not yet been detected and suddenly emerge in humans through cross-species transmission, or transfer from the atmosphere; the other is that endemic viruses undergo gene recombination, mutation to finally generate new pathogens. SARS-CoV and MERS-CoV are both presumably recombinants (Huang, Liu, Xu, et al., 2016; Sabir, Lam, Ahmed, et al., 2016), and our previous studies have shown that their emergence were somehow associated with double peaks in the sunspot cycle in 2002 and 2012 (Qu & Wickramasinghe, 2017). The Novel Coronavirus has been argued by some to have gained the ability to infect people through gene recombination and thus ultimately create a major public health threat on a world-wide scale (Ji, Wang, Zhao, et al., 2020).

According to the data of Sunspot Index and Long-Term Solar Observations, we are now approaching the deepest sunspot minimum for a century. There were 273 spotless days in 2019 and no sunspots for 40 days from November 14 to December 23, 2019. The Sun's magnetic field is very weak, thus allowing an increased flux of cosmic rays as well as charged dust particles (including viruses) to be incident on the Earth. Neutron counts from Moscow Neutron Monitor show that the flux of cosmic rays reaching Earth in 2019 is at a maximum over a timespan of half a century since 1962. It is of interest to note that immediately prior to the first recorded cases of the novel Corona virus in China a peak of cosmic rays was measured as is indicated by the Huon neutron monitor data in Fig. 1. Estimates of the timing of the most recent common ancestor of COVID-19 made with current sequence data point to emergence of the virus in late November 2019 to early December 2019, compatible with the earliest retrospectively confirmed cases (Huang et al., 2020) and the cosmic ray spike in late November 2019. Hence, we suggest that this strong cosmic ray spike was in some way connected with the onset of the outbreak.

A peak of cosmic rays occurring at a deep solar minimum not only shows the Earth's greater vulnerability to mutagenic effects on viruses already on the ground, but also through a weakening of the Earth's magnetic field environment, there would be an enhanced ingress of charged cosmic dust. In a series of papers (Wickramasinghe, Steele, Gorczynski, & Temple, 2020, 2020b, 2020c) the possibility of cometary dust laden with viruses as a

Fig. 1 Cosmic ray variations by the Huon neutron monitor data in China.

primary cause of the COVID pandemic has been discussed. In either case the mutagenic effect of cosmic rays could influence the course of the pandemic by affecting the evolution of the circulating virus after it is introduced into the human population.

Point mutations, gene recombination and gene reassortment are thought to be the three basic mechanisms of viral emergence and/or evolution. Solar radiation and cosmic rays are physical mutagens possibly leading to natural point mutations and can in this way lead to the emergence of new or modified viruses. Recombination and reassortment of viral genes occur at highly variable frequencies in viruses with RNA as their genetic material. Multiplicity reactivation and cross-reactivation have also been proven to be general mechanisms of genetic recombination. Such reactivation has been observed in influenza viruses irradiated by various types of radiations such as ultraviolet light and gamma rays, in laboratories during the1950s–60s. In a recent study we proposed that solar radiation and cosmic rays could lead to genetic mutation/recombination, and can thus cause the emergence of some modified viruses such as those responsible for pandemic influenza (Qu, 2016).

Cosmic rays can inactivate corona viruses, but the resulting seemingly noninfectious inactivated viruses are in fact semi-infectious. Semi-infectious virus particles deliver an incomplete set of viral genes to the cell that can, given a suitable trigger (cosmic rays), be provoked to become infective (Qu, 2016).

In a letter to *Current Science* published in November 2019, we also explicitly reminded the world that new infective viruses are likely to emerge in the coming months during the lowest minimum of the sunspot cycle in over a 100 years, and that public health authorities must be vigilant and take necessary action (Wickramasinghe, Wickramasinghe, Senanayake, et al., 2019). The emergence of the COVID-19, in our view, confirms this prediction and highlights the urgency of taking appropriate precautions without further prevarication.

Acknowledgments

Data on the number of cosmic rays were provided by Yang Ba Jing cosmic rays monitoring station of China. (http://ybjnm.ihep.ac.cn/munm/).

Competing interests

The authors have declared that no competing interests exist.

References

Andersen, K. G., Rambaut, A., Lipkin, W. I., et al. (2020). The proximal origin of SARS-CoV-2. *Nature Medicine*. https://doi.org/10.1038/s41591-020-0820-9.

Huang, C., Liu, W. J., Xu, W., et al. (2016). A bat-derived putative cross-family recombinant Coronavirus with a Reovirus gene. *PLoS Pathogens*, *12*(9), e1005883.

Huang, C., et al. (2020). Clinical features of patients infected with 2019 novel coronavirus in Wuhan, China. *Lancet*, *395*, 497–506. https://doi.org/10.1016/S0140-6736(20)30183-5.

Ji, W., Wang, W., Zhao, X., et al. (2020). Homologous recombination within the spike glycoprotein of the newly identified coronavirus may boost cross-species transmission from snake to human. *Journal of Medical Virology*, *92*(4), 433–440. https://doi.org/10.1002/jmv.25682.

Qu, J. (2016). Is sunspot activity a factor in influenza pandemics? *Reviews in Medical Virology*, *26*(5), 309–313.

Qu, J., & Wickramasinghe, N. C. (2017). SARS, MERS and the sunspot cycle. *Current Science*, *113*(8), 1501–1502.

Sabir, J., Lam, T., Ahmed, M., et al. (2016). Co-circulation of three camel coronavirus species and recombination of MERS-CoV in Saudi Arabia. *Science*, *351*(6268), 81–84.

Wickramasinghe, N. C., Steele, E. J., Gorczynski, R. M., & Temple, R. (2020b). Comments on the origin and spread of the 2019 coronavirus. *Virolology: Current Research*, *4*, 1.

Wickramasinghe, N. C., Steele, E. J., Gorczynski, R. M., & Temple, R. (2020c). "Growing evidence against global infection-driven by person-to-person transfer of COVID-19." *Virolology: Current Research*, *4*, 1.

Wickramasinghe, N. C., Steele, E. J., Gorczynski, R. M., Temple, R., et al. (2020). Predicting the future trajectory of COVID-19. *Virolology: Current Research*, *4*, 113.

Wickramasinghe, N. C., Wickramasinghe, N. T., Senanayake, S., et al. (2019). Space weather and pandemic warnings? *Current Science*, *117*(10), 232.

CHAPTER TEN

Microbial transfers from Venus to Earth

N. Chandra Wickramasinghe[a,b,c,d,*] and Predrag Slijepcevic[e]
[a]University of Buckingham, Buckingham, United Kingdom
[b]Centre for Astrobiology, University of Ruhuna, Matara, Sri Lanka
[c]National Institute of Fundamental Studies, Kandy, Sri Lanka
[d]Institute for the Study of Panspermia and Astroeconomics, Gifu, Japan
[e]Brunel University, London, United Kingdom
*Corresponding author: e-mail address: ncwick@gmail.com

Contents

Abstract

The possibility that the clouds of Venus are habitats for microorganisms has been discussed for several decades. Over the past two decades evidence to support this point of view has grown with new data from space probes and space exploration. In this article we argue that microorganisms are likely to be widely present in the clouds of Venus, and may under certain conditions have a ready route to Earth. Such transfers could occur by the action of the solar wind that leads to expulsion of parts of the atmosphere laden with microorganisms. The expelled material forms a comet-like tail in the antisolar direction and during inferior conjunctions of Venus could lead to injections of bacteria and other microorganisms onto the Earth. In situations of very low sunspot activity as now prevails, with a consequent weakening of the magnetopause this flux of microbes will be considerably enhanced. The inferior conjunction of 4 June 2020 together with the prevailing deep minimum in the sunspot cycle provides a combination of circumstances that is particularly favorable to such a process.

Advances in Genetics, Volume 106
ISSN 0065-2660
https://doi.org/10.1016/bs.adgen.2020.06.004

1. Introduction

A revival of interest in the transfer between planetary bodies of the solar system has followed the discovery of complex organic molecules in planets, moons, asteroids, comets and meteorites (Napier, 2004; Wallis & Wickramasinghe, 1995, 2004; Wickramasinghe, 2008). Most discussions have focussed thus far on the idea of impact-driven transfers of life-bearing rocks and dust that can take place sporadically over long dynamical timescales, $>10^7$ year. Transfers of microbiota over very much shorter timescales between neighboring planets are possible, but dynamical pathways of connection need to be identified. In this article we explore one such pathway for transfer of microorganisms between Venus and Earth. Such transfers may well have been taking place over several billion years, thus making these two terrestrial neighbors share a common microbial biosphere.

2. Clouds of Venus

Recently several studies have reviewed the possibility that the clouds of Venus could provide habitats for a variety of microbial extremophiles. Venus has a radius of about 95% of the Earth's radius and is at a distance of about 40 million km from Earth during inferior transit (closest approach). Conditions at the surface of Venus are distinctly inhospitable to life with temperatures averaging some 460 °C. About 30 miles (48 km) above the planet's surface, however, is a lower cloud layer with temperatures around 60 °C (140°F) and pressures similar to that on Earth. It is certainly not inconceivable that microbial life could exist in the clouds of Venus in this atmospheric Goldilocks zone. Here on Earth microorganisms have been found viable at altitudes as high as 25 miles (41 km) in the atmosphere, as well as at depths of 11 km in the Mariana Trench, Challenger Deep on the ocean floor (Kato, Lee, Nogi, et al., 1998).

The possibility of microbiology in the Venusian clouds has recently gained credence following discoveries made by the Japanese *Akatsuki* Venus Climate Orbiter that began returning results in December 2015 (Joseph, 2019; Limaye, Mogul, Smith, et al., 2018). Ultraviolet reflectance images showed dark patches that were comprised of sulfuric acid "particles" together with a population of unknown sub-micron-sized particles that absorb at an ultraviolet wavelength not characteristic of sulfuric acid. Indeed, it seems likely that atmospheric processes in the clouds of Venus

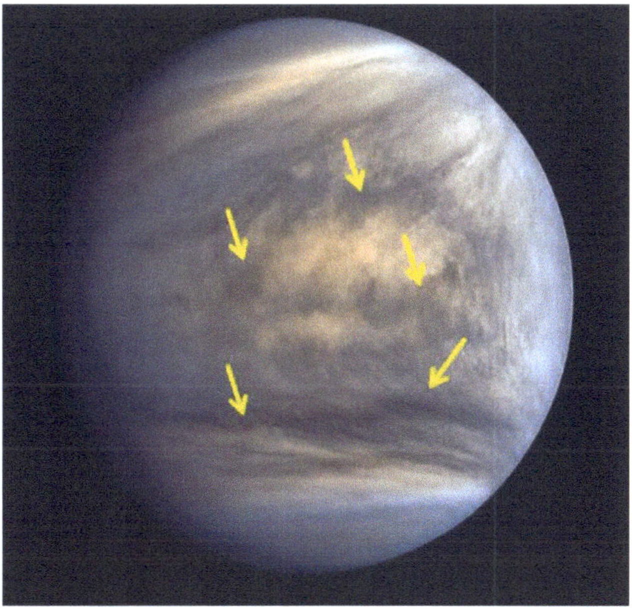

Fig. 1 Akatsuki ultraviolet image in false 2-color reduction. Cloud structure is due to unknown particulate component intimately involved in Venusian meteorology. *Adapted by the authors from images supplied by JAXA (Japan).*

are controlled and driven by microbial processes, microbiology then being in control of Venusian meteorology as Hoyle and one of us had suggested in 1979 (Fig. 1).

3. Exotic microbiology on Venus

Data from ESA's Venus Express probe (2007) provided evidence of frequent electrical discharges (lightning) in the atmosphere. Such energetic events would be expected to generate large amounts of CO from CO_2. The absence of higher concentrations of CO than is expected, despite the lightning, might be taken as strongly suggestive of an exotic microbiology. There is a diverse group of terrestrial bacteria and archaea known as hydrogenogens that can grow anaerobically using CO as the sole carbon source and H_2O as an electron acceptor, producing CO_2 and H_2 as waste products (Wu et al., 2005). The presence of H_2S and SO_2 in the atmosphere could also point to the presence of extremophilic "sulfur" bacteria (Cockell, 1999); and droplets of atmospheric sulfuric acid could provide a medium in which acidophiles can thrive. There is a case that can be made also for a variety of other microbial entities that populate the clouds of Venus, as has been reviewed by Limaye et al. (2018) and Joseph (2019).

Wilquet et al. (2012) studied the wavelength dependence of aerosol extinction in the clouds of Venus and inferred the existence of at least two types of particles: submicron haze particles with $r=0.1$–$0.3\,\mu m$ and larger particles with $r=0.4$–$1\,\mu m$. A model describing the upper haze on Venus leads to a bimodal distribution of particle size. This in our view could define the two classes of composition implied by the Akatsuki data—sulfuric acid droplets, and a UV-absorber/emitter of unknown origin leading to a changing configurations of dark patches, which in our view could most likely have a biological provenance.

The situation in Venus could be analogous to what happens in terrestrial clouds. Sattler, Puxhaum, and Psenner (2011) demonstrated bacterial growth in super-cooled cloud droplets, arguing that bacteria in tropospheric clouds are excellent sites for microbial replication. A stable aerobiology requires processes by which (a) bacteria nucleate droplets containing water and nutrients, (b) bacterial colonies grow within droplets, (c) the droplets fall into regions of higher temperature where they evaporate releasing spores to convect upward to produce further nucleation. In the case of Venus this cyclical feedback process would be expected to take place between the tops and bases of clouds, precisely the type of dynamic picture has emerged from the most recent data from the Japanese *Akatsuki* probe.

The latest data from the *Akatsuki* probe indeed confirms a multiplicity of microbial niches that could be present in the Venusian cloud system, which is perhaps the largest and most complex among all the terrestrial planets (Titov et al., 2018). A rapidly changing and complex pattern of cloud structure extending from 50 to 70 km and further out from the planet's surface as a diffuse dynamic haze is now amply documented (Limaye et al., 2018; Titov et al., 2018). The cloud opacity in the densest regions of cloud at visible and near-IR wavelengths could be exceedingly high at visible and near infrared wavelengths, showing remarkable spatial and latitudinal variability.

4. Solar wind erosion of the Venusian atmosphere

The effects of solar wind in producing erosion and evolution of planetary atmospheres over geological timescales have been extensively discussed and are now well recognized. However, on much shorter timescales, losses from the atmosphere due to solar wind interactions have in general tended to be ignored.

Atmospheric erosion due to the solar wind is expected to be most efficient for planets such as Venus that have negligible magnetic fields thus providing little or no confinement of charged particles. For this reason there have been early suggestions that Venus and comets may have similar characteristics in developing wakes or tails through interaction with the solar wind (Russell & Vaisberg, 1983; Russell et al., 1982; Wallis & Ip, 1982). While comet tails remain coherent over tens of millions of kilometers, the search for ions attributable to a Venusian origin beyond ~10 Venus radii had produced negative results until very recently (Kar, 1996) (Fig. 2).

The situation has recently changed dramatically, however. Firstly, Venus Express found evidence that the solar wind is causing a continual outflow of atmospheric gases including O^+ and H_2O^+ from the atmosphere of Venus in the form of a narrow wake or tail in the antisolar direction (Fig. 3) (Svedhem, Titov, Taylor, & Witasse, 2007). In view of the large-scale convective motions in the atmosphere, material in the wake would include material from the clouds where molecules and aerosols are present. Secondly, as part of the ion mass spectrometry program on the SOHO spacecraft, Grunwaldt et al. (1997) successfully detected a flux of O^+ and C^+ ions that originated in Venus and arrived near the Earth. The latter observation was made on June 10, when Venus was in inferior conjunction, but not in transit, and SOHO was 4.4×10^7 km downstream of Venus at the Earth's L1 Lagrangian point about 0.01 AU sunward of the Earth.

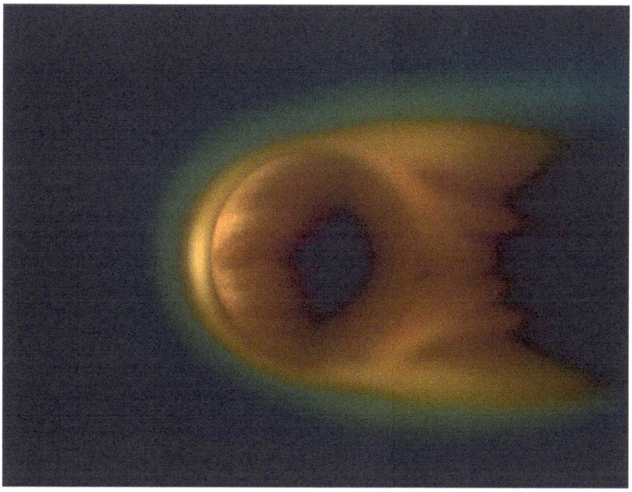

Fig. 2 Evidence of solar wind excavating the atmosphere of Venus *Courtesy ESA; http://www.esa.int/esaSC/SEMMAGK26DF_index_0.html; Venus Express data; H. Svedhem; image in public domain.*

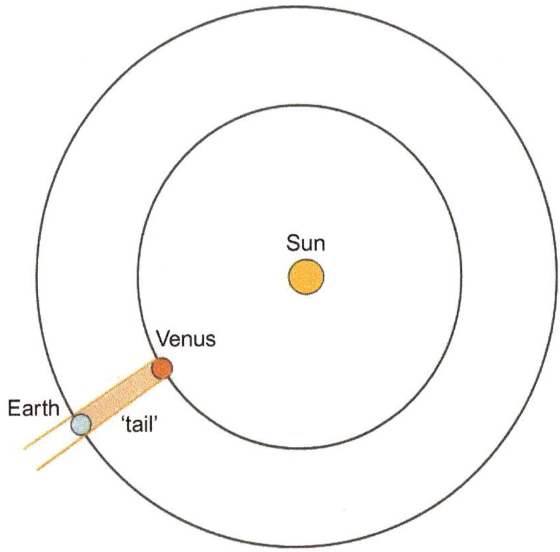

Fig. 3 Schematic mode of transfer of material from Venus to Earth during a planetary transit event. The tube of material reaching Earth is assumed to be 840 km wide, as indicated in the data of Grunwaldt et al. (1997).

Grunwaldt et al. (1997) estimated the O^+ flux originating from Venus to be $\sim 3 \times 10^3$ ions $cm^{-2}\, s^{-1}$ near the Earth. With a Venus ray (tail) diameter of 820 km Grunwaldt et al. obtained a total flux of O^+

$$F_{O^+} = 1.6 \times 10^{19} s^{-1} \quad (1)$$

with F_{C^+} being $\sim 0.1\ F_{O^+}$. The width of the Venusian tail as observed by Grunwaldt et al. at a heliocentric distance $r = 0.99$ AU will not be significantly different from that which could reach the Earth at $r = 1.0$ AU.

Since the orbit of Venus is inclined at 3.4° to that of the Earth, material transfer would most effectively take place during inferior conjunctions, and in particular (though not exclusively) those that occur along the line of nodes (transits of the planet Venus). In such situations we can estimate the mass reaching the Earth, based on the measurements of Grunwaldt et al. (1997). The geometry of the Venus to Earth transfer process envisaged here is schematically shown in Fig. 2.

If ω_V and ω_E denote the angular velocities of Venus and Earth respectively in their orbits, the velocity relative to Earth of the footprint of the Venusian tail (assumed to be a 840 km wide cylinder) is

$$v = a_E(\omega_V - \omega_E) = 2\pi\left(a_V^{-3/2} - 1\right) \text{AU/year} \quad (2)$$

using Kepler's law. Setting $a_v = 0.75$ AU, we obtain a footprint speed of the Venusian tail to be

$$v \approx 20\,\text{km/s} \tag{3}$$

The effective trapping cross-section for particles with the Earth will be larger than its geometrical cross-section because they are charged. It would be reasonable to assume that the trapping path length is the dimension of the magnetopause, say $\sim 20\ R_E \sim 1.3 \times 10^5$ km. Using (Eq. 3) the inoculation time of Venusian particles into the Earth's environment would therefore be ~ 6000 s.

From Eq. (1) the total number of O^+ ions introduced into the Earth's magnetosphere during each encounter is $\sim 10^{23}$, with the number of C^+ ions being a factor ~ 10 less (Grunwaldt et al., 1997).

The solar wind at Venus could be assumed to have excavated not only CO_2 (which dissociates into C^+ and O^+) but also dust which forms a substantial fraction of the middle and upper atmosphere. Dust grains (including bacteria) would be charged and coupled to the outflowing plasma tail. It would be entirely reasonable to speculate that the mass of grains carried along with the ions makes up $\sim 1\%$ of the total outflow. Thus the mass of bacterial material carried into the Earth environment along with the O^+ and C^+ during a transit amounts to

$$M_{\text{bact}} \approx 0.03\ \text{g}. \tag{4}$$

The dry mass of a typical bacterium being $\sim 10^{-13}$ g, the total number of bacteria entering the Earth's magnetosphere is then $\sim 10^{11}$.

A possible transfer of $\sim 10^{11}$ microbial cells from Venus to the Earth's environment in a single injection event, although modest, cannot be ignored. It would require only a small fraction of this inoculant to find an appropriate terrestrial niche to effectively link the biosphere of Venus to that of the Earth. However, it should be noted that the estimate $\sim 10^{11}$ bacteria, based on a single set of measurements made by Grunwaldt et al., is on the conservative side, and possibly represents a lower limit. The time in 1997 when Grunwaldt et al. detected ions in the Venus tail near the Earth coincided with a deep minimum in the sunspot cycle. It is well-known that solar flare and solar wind activity increase near sunspot maximum, and varies erratically from one cycle to the next. Whenever a transit occurs during an active phase of the sun the bacterial mass from Venus that is added to the Earth could well exceed 10 g, giving rise to a total inoculant containing

$\sim 10^{14}$ bacteria. Survival of microbes (particularly radiation resistant microbes) in traveling between Venus and Earth would not pose a problem. The time of exposure to solar radiation in unshielded interplanetary space is $\sim 10^5$ s, so high survival rates are to be expected under the most pessimistic assumptions (Mileikowsky et al., 2000).

The fate of charged bacteria entering the magnetopause of the Earth is probably well represented by the known behavior of 0.1–1 μm sized particles from Lunar ejecta that are known to enter the Earth (Mueller & Kessler, 1985; Zook et al., 1984). Acted upon by the charge-dependent Lorentz force, convective drag forces, solar radiation pressure and gravity, bacterial dust grains diffuse through the magnetosphere to reach the atmosphere on timescales of ~20–200 h. However, the details of the entry process involve uncertainties and complications arising from charging and discharging of grains, as well as instabilities in the field configurations of the magnetosphere. It is reasonable nevertheless to assume that a large fraction of the particles find their way ultimately into the troposphere where they could serve as condensation nuclei of raindrops. If the terrestrial clouds can offer a niche that matches the essential characteristics of the Venusian clouds, an alien aerobiology would become established on Earth. Otherwise, the iterant Venusian microbes are discharged in the rain to either find a perhaps implausible surface niche, or to perish.

5. Early work of D.R. Barber

As far as we can see the idea microbes from Venus reaching Earth was first proposed by Donald Barber (1963) in an attempt to explain a periodic "fogging" of photographic plates that he found to take place at Sidmouth Observatory. Over the period from 1937 to 1961 Barber reported six incidents of bacteria causing the fogging of photographic plates each of which occurred close (within a few weeks) to a date when there was an inferior conjunction of the planet Venus. Barber attributed this to a bacterium from Venus that was presumably introduced into the rain water used for the developing of photographic plates. At the time astrobiology and the possibility of microbial life in the clouds of Venus would have sounded as bizarre as the stories of fairies in the garden, and consequently these observations went unnoticed. In the context of the arguments presented in this review the interpretation offered by Barber of the strange events at Sidmouth Observatory may not seem so outrageous after all. The Earth and her sister planet Venus appear to be intimately connected. Although conditions at the surface of Venus is at present truly Hadean in regard to biological habitability, the existence of friendly surface conditions less than a billion years ago is considered to be a distinct possibility.

6. Future inferior conjunctions and microbial transfers

The next three inferior conjunctions of Venus when transfers of microbial material to the vicinity of the Earth's orbit could take place are:

June 3, 2020
January 9, 2022
August 13, 2023

None of these coincide with transits so the most direct transfers may not be thought possible. However, in an extended wake of expelled plasma which is expected, the intersection with the Earth can by no means be excluded. Indeed the time when Grunwaldt et al. made their oxygen ion detection was an inferior conjunction that was *not* a transit of Venus (Fig. 4).

Another circumstance to consider in the forthcoming conjunction is that we are currently experiencing the deepest sunspot minimum for over a century. So the ingress of charged microorganisms, which may be carried within a diverging tube of plasma emanating from Venus, would be greatly

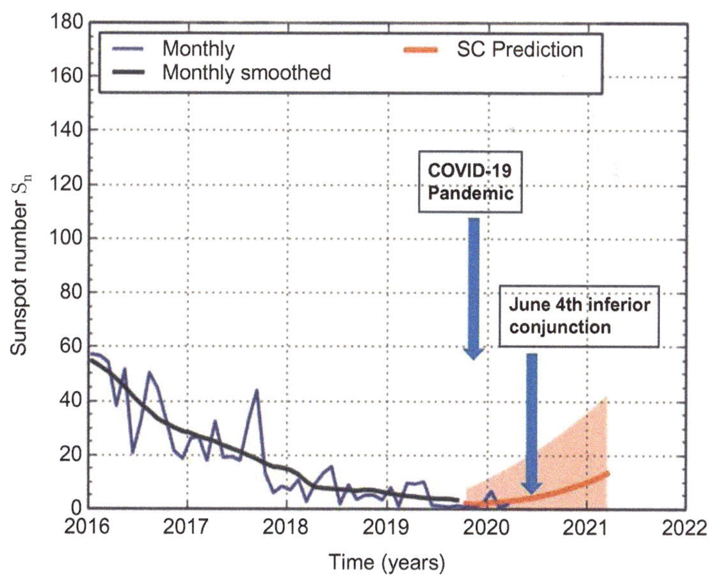

Fig. 4 Sunspot cycle showing lowest minimum in over 100 years. *Adapted from data provided by the Royal Observatory of Belgium. (author's copyright).*

facilitated due through a weakened magnetopause. Whether we might expect new bacterial species to emerge on the Earth some months after June 4, 2020 (perhaps similar to *Candia auris*, discussed in another chapter, which arrived under similar conditions in 2007) remains to be seen. However, the microbiological connectivity between Venus and Mars has now become more real and urgent than it has ever been before.

References

Barber, D. (1963). Perspective. In (Vol. 5), London: Focal Press, pp. 201–208.

Cockell, C. S. (1999). *Planetary and Space Science, 47*, 1487.

Grunwaldt, H., et al. (1997). *Geophysical Research Letters, 24*, 1163.

Joseph, R. G. (2019). Life on Venus and the interplanetary transfer of biota from Earth. *Astrophysics and Space Science, 364*, 191.

Kar, J. (1996). *Space Science Reviews*, 77, 193.

Kato, C., Lee, L., Nogi, Y., et al. (1998). Extremely barophilic bacteria isolated from the Mariana trench, challenger deep, at a depth of 11,000 meters. *Applied and Environmental Microbiology, 64*(4), 1510–1515.

Limaye, S. S., Mogul, R., Smith, D. J., et al. (2018). Venus' spectral signatures and the potential for life in the clouds. *Astrobiology, 18*(9), 1181–1198. https://doi.org/10.1089/ast.2017.1783.

Mileikowsky, C., et al. (2000). *Icarus, 145*, 391.

Mueller, A., & Kessler, D. (1985). *Advances in Space Research, 5*, 77.

Napier, W. M. (2004). *MNRAS, 348*, 46.

Russell, C. T., & Vaisberg, O. (1983). In D. M. Huntenet al. (Ed.) *Venus* (pp. 873–940). Univ. Ariz. Press.

Russell, C. T., et al. (1982). In L. L. Wilkening (Ed.), *Comets* (p. 561). Univ. Aris Press.

Sattler, B., Puxhaum, H., & Psenner, R. (2011). *Geophysical Research Letters, 28*, 239.

Svedhem, H., Titov, D. V., Taylor, F. W., & Witasse, O. (2007). *Nature, 450*, 629.

Titov, D. V., et al. (2018). Clouds and hazes of Venus. *Space Science Reviews, 214*, 126.

Venus Express (ESA), 2007, http://www.esa.int/esaSC/SEMMAGK26DF_index_0.html

Wallis, M. K., & Ip, W.-H. (1982). *Nature, 298*, 229.

Wallis, M. K., & Wickramasinghe, N. C. (1995). *Earth and Planetary Science Letters, 130*, 69.

Wallis, M. K., & Wickramasinghe, N. C. (2004). *MNRAS, 348*, 52.

Wickramasinghe, J. T. (2008). *The role of comets in panspermia.* PhD thesis, Cardiff University.

Wilquet, V., et al. (2012). Optical extinction due to aerosols in the upper haze of Venus: Four years of SOIR/VEX observations from 2006 to 2010. *Icarus, 2017*(2), 875–881.

Wu, M., et al. (2005). *PLoS Genetics, 1*(5), e65.

Zook, H., et al. (1984). *Lunar and Planetary Science, XV*, 965.

CHAPTER ELEVEN

Experiments to prove continuing microbial ingress from Space to Earth

N. Chandra Wickramasinghe[a,b,c,d,*], Edward J. Steele[a,e,f,*], Robert Temple[g], Gensuke Tokoro[b], Willam A. Smith[b], Brig Klyce[h], Dayal T. Wickramasinghe[i], and Dhammika Magana Arachchi[d]

[a]Centre for Astrobiology, University of Ruhuna, Matara, Sri Lanka
[b]Institute for the Study of Panspermia and Astroeconomics, Gifu, Japan
[c]University of Buckingham, Buckingham, United Kingdom
[d]National Institute of Fundamental Studies, Kandy, Sri Lanka
[e]C.Y.O'Connor ERADE Village Foundation, Piara Waters, Perth, WA, Australia
[f]Melville Analytics Pty Ltd, Melbourne, VIC, Australia
[g]The History of Chinese Science and Culture Foundation, Conway Hall, London, United Kingdom
[h]Astrobiology Research Trust, Memphis, TN, United States
[i]College of Physical and Mathematical Sciences, Australian National University, Canberra, ACT, Australia
*Corresponding authors: e-mail address: ncwick@gmail.com; e.j.steele@bigpond.com

Contents

Abstract

A wide range of evidence for pointing to our cosmic origins is close to the point of being overwhelming. Yet the long-entrenched paradigm of Earth-centered biology appears to prevail in scientific culture. A matter of crucial importance is to carry out a decisive experiment that is long overdue—establishing empirically beyond any doubt that extraterrestrial microbiota reaches the surface of the Earth at the present day. Such an experiment may of course happen naturally by the appearance of pandemics of new disease as discussed in an earlier chapter.

Advances in Genetics, Volume 106
ISSN 0065-2660
https://doi.org/10.1016/bs.adgen.2020.03.006

1. Early evidence

An obvious place to discover biological material from comets is in the Earth's upper atmosphere itself. Cometary meteoroids (fragments of comets) and interplanetary and interstellar dust particles are known to enter the atmosphere in vast quantity, at a more or less steady rate, averaging in excess of $\sim 10^2$ ton/day. Although much of this incoming material burns up as meteors, a significant fraction survives entry. Organic grains including bacteria and viruses of micrometer sizes, arriving as clumps and dispersing in the high stratosphere, would be slowed down gently and would not be destructively heated. The Earth's atmosphere could thus serve as an ideal collector of biological material and the potential for both random pandemics and further evolutionary progress.

The Australian scientist Bigg (1983) was among the first to actually recover stratospheric particles from above 25 km using balloons. His collections dating from 1960 onward clearly reveal the presence of oval shaped dust particles, and particles endowed with fimbrai, which Bigg himself speculated may have a possible biological provenance. A few years later Greene, Hagberg, Lundgren, and Pederson (1965) carried out a series of balloon flights to recover microorganisms in the two height ranges 30,000–60,000 ft. and 60,000–90,000 ft. in the stratosphere. Positive results in both these height ranges possibly pointed to an in-falling microbial population, and this came as an unwelcome surprise in the years running up to the dawn of the space age. Although some of the microorganisms collected in early experiments were claimed to exhibit "unusual" properties, such as pigmentation and radiation resistance, their possible extraterrestrial origin remained a matter of dispute. No DNA sequencing procedure was available at the time to ascertain significant deviations there might have been from any related terrestrial species. Furthermore, the collection and laboratory techniques available at the time left open a high chance of contamination. A history of early attempts to probe the stratosphere for microorganisms is summarized by Smith (2013).

2. Millennial studies

Methods for the collection of cometary microbiota in the stratosphere, if they are to be decisive, should involve procedures for either sifting out a terrestrial component or for excluding contamination altogether. In 2001 the Indian Space Research Organization (ISRO) launched a balloon into the stratosphere to heights between 20 and 41 km to collect stratospheric air under aseptic conditions with a view to testing the theory of cometary panspermia (Harris et al., 2001). The procedure involved the use of cryogenically cooled, stainless steel cylinders that were evacuated and autoclaved and fitted with valves that could be opened at predetermined altitudes (Fig. 1). A large quantity of stratospheric air at 41 km was collected in this manner and the cylinders containing this material parachuted back for laboratory study.

The air within the pressurized cylinders was carefully released and passed through millipore membrane filters under carefully controlled laboratory conditions. Aerosols (dust and bacteria) present within the volume of stratospheric air were thus trapped and collected on the filters. The particles recovered fell into two broad categories: (a) mineral grain aggregates, very similar to Brownlee particles (inorganic interplanetary dust) and (b) fluffy carbonaceous aggregates which could be identified as clumps of bacteria (see Fig. 2A). Typical dimensions of the clumps were about 10 μm.

Fig. 1 (A) Cylinder containing stratospheric air and (B) stack of cylinders launched within a liquid neon container. *From Wickramasinghe with permission.*

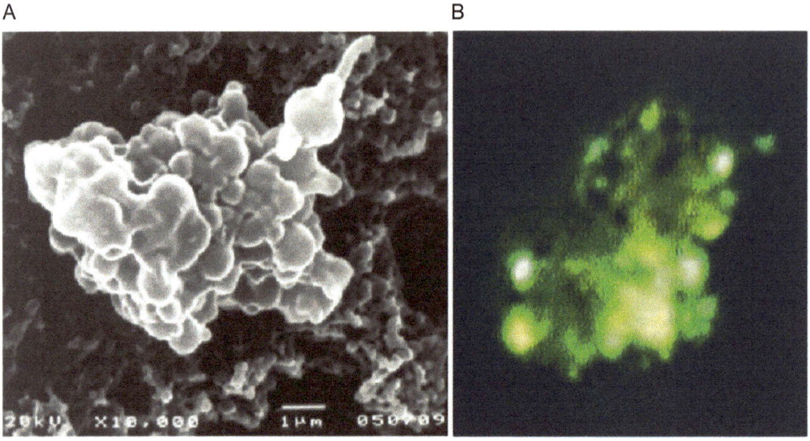

Fig. 2 (A) A carbonaceous stratospheric particle from 41 km resembling a clump cocci and a rod bacterium. Panel (B) A clump of viable but non culturable bacteria flourescing in carbocyanine dye.

In addition to structures such as shown in Fig. 2A, which were presumed to be bacteria, the stratospheric samples from 2001 also revealed evidence of similar-sized bacterial clumps that could not be cultured but were detected by the use of a fluorescent dye (carbocyanine dye). The uptake of the dye revealed the presence of living cells in the clumps (see Fig. 2B) thus interpreted to be "Viable but non-culturable microorganisms." The cometary origin of all such particles was inferred on the grounds that the altitude of 41 km was too high for lofting 10-μm sized clumps of solid material from the Earth's surface under normal conditions.

3. Flux of microbes from space

It was next possible to infer the in-fall rate of biological particles at 41 km from the data obtained in the balloon launch of 2001 (Harris et al., 2001). When the high pressure stratospheric air within the recovered cylinders was passed through Millipore filters Harris et al. (2001) discovered an average of one clump of putative biological cells per 25 mm^2 of filter area. With a total membrane filter area of 2000 mm^2 and a total NTP volume of air released being measured as the entire membrane filter would have contained ~80 clumps of cells that must therefore represent the bacterial component of the air within the cylinder that was collected from ~41 km. With volume of air released being recorded as 18.5 L the NTP concentration of

Table 1 Falling speed w (cm/s).

***a*(micron)** z(km)	0.3	1	3	10
40	0.12	0.36	1.52	5.37
20	0.0066	0.036	0.187	1.69
10	0.0019	0.017	0.137	1.52

clumps is thus ~80/18.5 = 4.3 per liter. Converting this to the ambient conditions at the point of collection ~39–41 km ($P = 2.9 \times 10^{-3}$ bar, $T = 253$ K) we obtain a local clump density of 1.4×10^{-2} per liter of ambient stratospheric air at 40 km altitude.

Next, we need to estimate the falling (terminal) speed of aerosols falling through the stratosphere at various heights. For this we use data set out by Kasten (1968) confirmed by later studies of Colbeck (1998). From Kasten's figures we can calculate the falling speeds of spherical particles of various radii and unit specific gravity as shown in Table 1:

The average mass for a 3 μm radius clump can be taken as $\sim 3 \times 10^{-11}$ g, and the settling speed of such a clump at 40 km 1.52 cm/s according to Table 1 is ~1.52 cm/s. For the entire Earth's surface of 5×10^{15} cm^2, we thus calculate an in-fall rate.

$$\sim 1.4 \times 10^{-5} \times 3 \times 10^{-11} \times 5 \times 10^{18} \times 1.52 = 3.19 \times 10\,\text{g}\,\text{s}^{-1}$$
$$\simeq 3\,\text{ton per day.}$$

This converts to some 10^8 bacteria per square meter arriving from space at the Earth's surface every day, along with possibly 10 times as many associated viruses.

4. Experiments by Wainwright's team

From 2013 onward Milton Wainwright and his team conducted a series of balloon flights in Chester, England to collect microbes from heights in the range 21–25 km in the stratosphere (Wainwright, Rose, Baker, Wickramasinghe, Omairi et al., 2015; Wainwright, Wickramasinghe, Harris, & Omairi, 2015; Wainwright, Wickramasinghe, Narlikar, & Rajaratnam, 2003; Wainwright, Wickramasinghe, Rose, & Baker, 2014). The sampler included a drawer mechanism that could be opened and closed at a predetermined height and new electron microscope stubs were exposed to falling microbes. The particles fell at speeds high enough to crater some of the carbon stubs, and many particles were discovered to be embedded in the stubs when they were recovered. The stubs were examined in the laboratory using techniques of electron microscopy. With controls taken to avoid contamination at every stage Wainwright et al. (2003) discovered 10 μm organic structures to which they ascribed a biological provenance.

5. Experiments planned by NIFS Sri Lanka

A group of us are planning to launch a balloon to 41 km over central Sri Lanka in mid-2020 for collecting stratospheric aerosols. The collected material would be examined using a variety of methods including *in situ* DNA/RNA analysis, and a possible non-terrestrial component will be identified in this way (program reviewed briefly in Steele et al., 2019).

6. Bacteria and viruses in the earth environment

It is only relatively recently that scientists have been able to fully grasp the enormous magnitude of the microbial and viral content of the terrestrial biosphere. We now know that a typical liter of surface seawater contains at least 10 billion microbes as well as some 100 billion viruses—the vast majority of which remain unidentified and uncharacterized to date (Weitz & Wilhelm, 2013). Two years ago an international group of scientists collected bacteria and viruses that fell through the rarefied atmosphere near the 4000 m peaks of the Sierra Nevada mountains of Spain. They arrived at an astonishing tally of some 800 million viruses per square meter per day and an associated slightly smaller tally of bacteria—all of which would of course ultimately fall to the Earth's surface (Reche, D'Orta, Mladenov, Winget, & Suttle, 2018). The assumption normally made is that all such viruses and bacteria necessarily originate on the Earth's surface and are swept upwards in air currents; but in such a model many difficulties associated with the upward transport process are ignored. In our view, a significant fraction of this vast number of falling microbes must originate outside the terrestrial biosphere and come from cometary sources—viruses and bacteria that are actually expelled from comets.

If we take into account all the facts available to date we cannot avoid the conclusion that vast numbers of bacteria and viruses continue to fall through the Earth's atmosphere, and it seems inevitable that a significant fraction is of external origin.

Comets have been regarded with awe and trepidation in many ancient cultures of the world. Almost without exception they have been regarded as bad omens—bringers of pestilence and death. The evidence for comets being implicated in the origin of life on Earth was intensely controversial when these ideas were first discussed by one of us and the late Sir Fred Hoyle. Now there is a growing consensus that this is inevitable in some form. In this article we argue that even today the periodic influx of cometary dust and debris could be responsible for waves of epidemic disease—such as the recent coronavirus (see Chapter "Origin of New Emergent Coronavirus and Candida Fungal Diseases – Terrestrial or Cosmic?" by Edward Steele et al.)—that sweep our planet from time to time.

As a life-bearing comet makes its repeated orbits around the sun its volatile substances are progressively vaporized and eventually we end up with what could be recognized as large carbonaceous meteorites. The number of close perihelion passages that a comet can survive before becoming completely stripped of volatiles is probably a few hundred. Carbonaceous chondrites could represent materials from comets denuded of volatiles but retaining a residue of silicates and more refractory organic structures. From time to time such objects find ingress into the Earth.

7. Direct evidence of microbial fossils in meteorites

Organic structures identifiable with bacteria, eukaryotic cells, and viruses have been reported in carbonaceous meteorites over several decades, including studies by Pflug in the 1980, and Richard Hoover in more recent times. Fig. 3 shows carbonaceous structures in the Murchison meteorite identified with fossilized microbiota.

The morphological resemblance of the virus-like carbonaceous structure identified by Plug is astoundingly similar to the coronavirus (Fig. 4). In 2013 a fall of carbonaceous stones in Sri Lanka also revealed distinctly biological structures as shown in Fig. 5. However, because this latter fall occurred over a remote part of the island and no video record was available, this event has not yet been registered as an official meteorite fall.

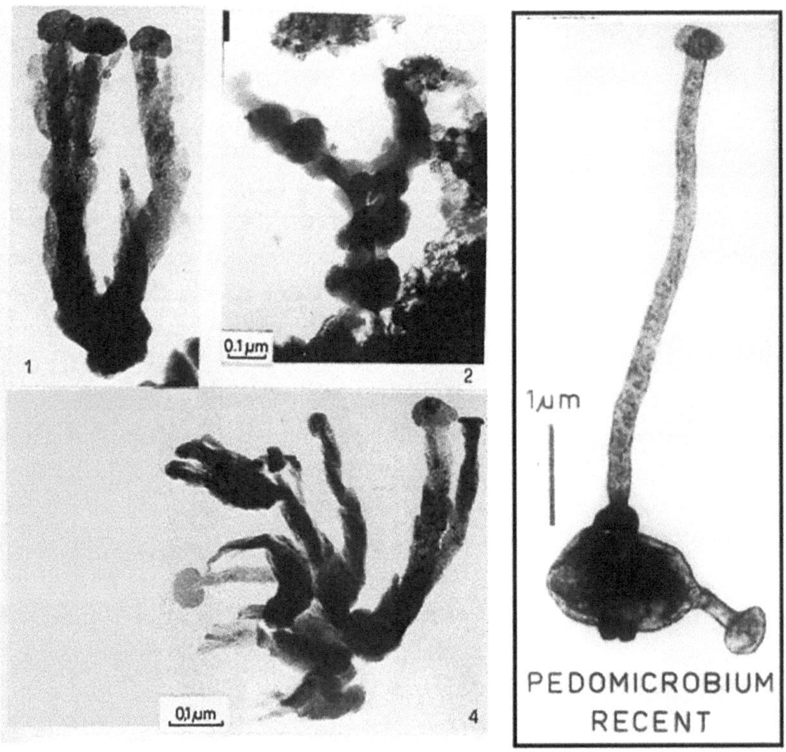

Fig. 3 Microfossils in the Murchison meteorite (left) discovered by Plug (1984). *Courtesy Pflug with permission Wickramasinghe.*

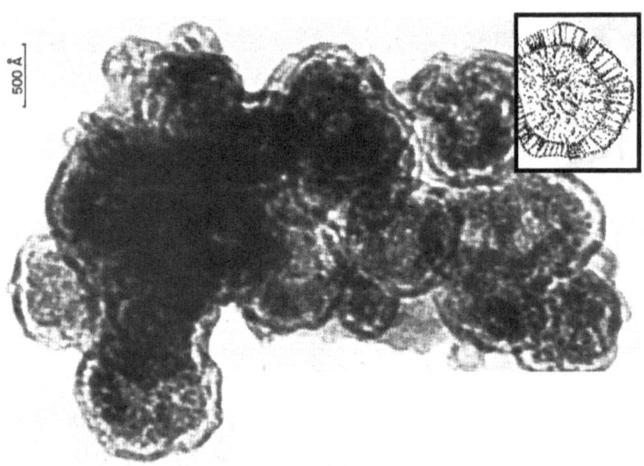

Fig. 4 Electron micrograph of organic structure within the Murchison meteorite compared with sketch of the structure of an influenza virus. *Published in Wickramasinghe, N. C., Steele, E. J., Gorczynski, R. M., Temple, R., Tokoro, G., Klyce, B., et al. (2020). Comments on the origin and spread of the 2019 coronavirus.* Virology: Current Research 4, *1. doi: 10.37421/Virol Curr Res.2020.4.435.*

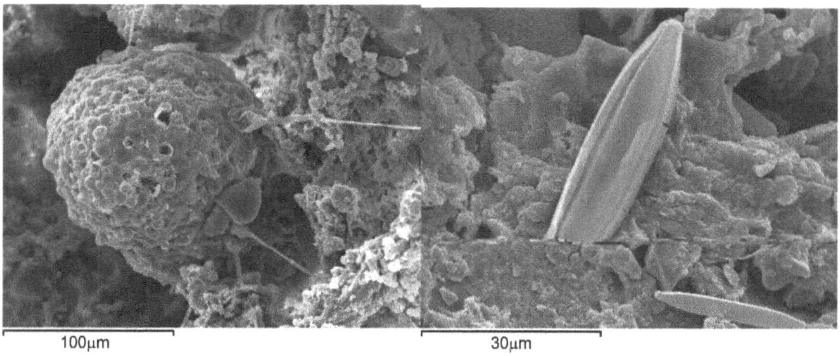

Fig. 5 Extinct acritarch fossil (L) and diatom frustule (R) from Polonnaruwa meteorite. C.f. Wallis et al. (2013) and Wickramasinghe, Wallis, Wallis, and Samaranayake (2013).

8. Problem of distinguishing indigenous terrestrial microbiota from bacteria of space origin

First and foremost, it will be important to repeat these early attempts to isolate microorganisms from the stratosphere and possibly beyond. In this context it is worth noting that Grebennikova, Syroeshkin, Shubralova, et al. (2018) have already isolated microorganisms above the stratosphere on the windows of the International Space Station at 400 km and Wickramasinghe and Rycroft (2018) have discussed the difficulties of lofting such particles to reach these heights.

Microbial entities collected in the stratosphere (or above) can be analyzed in many ways to establish their space origin. The use of nanosims technology to determine isotopic signatures could be one route that was attempted in the case of the 2001 recoveries without success, but these studies should be continued more intensively. The fact that appropriate facilities are few and far between on the planet, and access is expensive and difficult is a limitation that needs to be overcome.

If both the space incident microorganisms and terrestrial microbes originate from disconnected pieces of a single cosmic biosphere (Earth and space) their genetic difference may well turn out to be subtle and even difficult to detect. Indeed, an ISRO sponsored balloon flight into the stratosphere in 2006 recovered three new bacterial species that are genetically similar (80% homologous) to known terrestrial species but sufficiently different to be classified as different species (Shivaji, Chaturvedi, Begum, et al.,

2009). The first of the new species recovered from 41 km was named *Janibacter hoylei,* after Fred Hoyle.

To conclude, we would remark that signs of imminent change—a major paradigm shift—have come from many directions and such a paradigm shift is now unavoidable (Steele et al., 2018, 2019). The crucial data to clinch this shift must come from the results of experiments such as we have discussed here involving the recovery of microbes of space origin, and by establishing such an origin beyond a shadow of doubt. In this way we would conclude that the evolution of life takes place not just within a closed biosphere on our minuscule planet Earth but extends over a vast and connected volume of the cosmos.

References

Bigg, E. K. (1983). Particles in the upper atmosphere. In C. Wickramasinghe (Ed.), *Fundamental studies and the future of science* (pp. 38–51). University College, Cardiff Press.

Colbeck, I. (1998). Introduction to aerosol science. In I. Colbeck (Ed.), *Physical and chemical properties of aerosols*. London: Blackie Academic & Professional.

Grebennikova, T. V., Syroeshkin, A. V., Shubralova, E. V., et al. (2018). The DNA of bacteria of the world ocean and the Earth in cosmic dust at the international space station. *Scientific World Journal, 2018*, 7360147, Published online 2018 Apr 18. https://doi.org/10.1155/2018/7360147, https://www.hindawi.com/journals/tswj/aip/7360147/.

Greene, V., Hagberg, C., Lundgren, D., & Pederson, P. (1965). Microbiological exploration of stratosphere: Results of six ex-perimental flights. In *Proceedings of the atmospheric biology conference, University of Minnesota, 13–15 April 1964, NASA, Washington, DC* (pp. 199–212).

Harris, M. J., Wickramasinghe, N. C., Lloyd, D., Narlikar, J. V., Rajaratnam, P., Turner, M. P., et al. (2001). The detection of living cells in stratospheric samples. *Proceedings of SPIE, 4495*, 192–198.

Kasten, F. (1968). Falling speed of aerosols. *Journal of Applied Meteorology*, 7, 944–947.

Plug, H. D. (1984). In N. C. Wickramasinghe (Ed.), *Fundamental studies and the future of science*. University College Cardiff Press.

Reche, I., D'Orta, G., Mladenov, N., Winget, D. M., & Suttle, C. A. (2018). Deposition rates of viruses and bacteria above the atmospheric boundary layer. *The ISME Journal, 12*, 1154–1162. https://doi.org/10.1038/s41396-017-0042-4.

Shivaji, S., Chaturvedi, P., Begum, Z., et al. (2009). Janibacter hoylei sp. nov., Bacillus isronensis sp. nov. and Bacillus aryabhattai sp. nov., isolated from cryotubes used for collecting air from the upper atmosphere. *International Journal of Systematic and Evolutionary Microbiology, 59*, 2977–2986.

Smith, D. J. (2013). Microbes in the upper atmosphere and unique opportunities for astrobiology research. *Astrobiology, 13*(20), 981–990. https://doi.org/10.1089/ast.2013.1074.

Steele, E. J., Al-Mufti, S., Augustyn, K. A., Chandrajith, R., Coghlan, J. P., Coulson, S. G., et al. (2018). Cause of Cambrian explosion—Terrestrial or cosmic? *Progress in Biophysics and Molecular Biology, 136*, 3–23. https://doi.org/10.1016/j.pbiomolbio.2018.03.004.

Steele, E. J., Gorczynski, R. M., Lindley, R. A., Liu, Y., Temple, R., Tokoro, G., et al. (2019). Lamarck and Panspermia—On the efficient spread of living systems throughout the cosmos. *Progress in Biophysics and Molecular Biology, 149*, 10–32, pii: S0079-6107(19)30112-9. https://doi.org/10.1016/j.pbiomolbio.2019.08.010.

Wainwright, M., Rose, C. E., Baker, A. J., Wickramasinghe, N. C., Omairi, T., et al. (2015). Biological entities isolated from two stratosphere launches-continued evidence for a space origin. *Journal of Astrobiology & Outreach*, *3*(2). https://doi.org/10.4172/2332-2519.1000129.

Wainwright, M., Wickramasinghe, N. C., Harris, M., & Omairi, T. (2015). Masses staining positive for DNA-isolated from the stratosphere at a height of 41 km. *Journal of Astrobiology & Outreach*, *3*(2). https://doi.org/10.4172/2332-2519.1000130.

Wainwright, M., Wickramasinghe, N., Narlikar, J., & Rajaratnam, P. (2003). Microorganisms cultured from stratospheric air samples obtained at 41km. *FEMS Microbiology Letters*, *218*, 161–165.

Wainwright, M., Wickramasinghe, N. C., Rose, C. E., & Baker, A. J. (2014). Recovery of cometary microorganisms from the stratosphere. *Journal of Astrobiology & Outreach*, *2*(1). https://doi.org/10.4172/2332-2519.1000110.

Wallis, J., Miyake, N., Hoover, R. B., Oldroyd, A., Wallis, D. H., Samaranayake, A., et al. (2013). The Polonnaruwa meteorite; Oxygen isotope, crystalline and biological composition. *Journal of Cosmology*, *22*(2), 10004–10011. published, 5 March 2013.

Weitz, J. S., & Wilhelm, S. W. (2013). An ocean of virsuses. *The Scientist* June 30, 2013. https://www.the-scientist.com/features/an-ocean-of-viruses-39112?archived_content=9BmGYHLCH6vLGNdd9YzYFAqV8S3Xw3L5.

Wickramasinghe, N. C., Wallis, J., Wallis, D. H., & Samaranayake, A. (2013). Fossil diatoms in a new carbonaceous meteorite. *Journal of Cosmology*, *21*(37), arXiv:1303.2398. published, 10 January 2013.

Wickramasinghe, N. C., & Rycroft, M. J. (2018). On the difficulty of the transport of electrically charged submicron dust from the Earth's surface to the high ionosphere. *Advances in Astrophysics*, *3*, 150–153. August 2018. 10.22606/adap.2018.33003.

FSC
www.fsc.org
MIX
Papier | Fördert
gute Waldnutzung
FSC® C083411

Kolibri 360 GmbH
Ferdinand-Jühlke-Straße 7
99095 Erfurt, Deutschland
produktsicherheit@kolibri360.de